Chauncey Grätzel

MEMS and High Speed Vision

Chauncey Grätzel

# MEMS and High Speed Vision

## Development and Application to Reverse-engineer Drosophila Flight Control

Südwestdeutscher Verlag für Hochschulschriften

**Impressum/Imprint (nur für Deutschland/ only for Germany)**
Bibliografische Information der Deutschen Nationalbibliothek: Die Deutsche Nationalbibliothek verzeichnet diese Publikation in der Deutschen Nationalbibliografie; detaillierte bibliografische Daten sind im Internet über http://dnb.d-nb.de abrufbar.
Alle in diesem Buch genannten Marken und Produktnamen unterliegen warenzeichen-, marken- oder patentrechtlichem Schutz bzw. sind Warenzeichen oder eingetragene Warenzeichen der jeweiligen Inhaber. Die Wiedergabe von Marken, Produktnamen, Gebrauchsnamen, Handelsnamen, Warenbezeichnungen u.s.w. in diesem Werk berechtigt auch ohne besondere Kennzeichnung nicht zu der Annahme, dass solche Namen im Sinne der Warenzeichen- und Markenschutzgesetzgebung als frei zu betrachten wären und daher von jedermann benutzt werden dürften.

Verlag: Südwestdeutscher Verlag für Hochschulschriften Aktiengesellschaft & Co. KG
Dudweiler Landstr. 99, 66123 Saarbrücken, Deutschland
Telefon +49 681 37 20 271-1, Telefax +49 681 37 20 271-0, Email: info@svh-verlag.de
Zugl.: Zürich, ETHZ, Diss., 2008

Herstellung in Deutschland:
Schaltungsdienst Lange o.H.G., Berlin
Books on Demand GmbH, Norderstedt
Reha GmbH, Saarbrücken
Amazon Distribution GmbH, Leipzig
ISBN: 978-3-8381-0663-2

**Imprint (only for USA, GB)**
Bibliographic information published by the Deutsche Nationalbibliothek: The Deutsche Nationalbibliothek lists this publication in the Deutsche Nationalbibliografie; detailed bibliographic data are available in the Internet at http://dnb.d-nb.de.
Any brand names and product names mentioned in this book are subject to trademark, brand or patent protection and are trademarks or registered trademarks of their respective holders. The use of brand names, product names, common names, trade names, product descriptions etc. even without a particular marking in this works is in no way to be construed to mean that such names may be regarded as unrestricted in respect of trademark and brand protection legislation and could thus be used by anyone.

Publisher:
Südwestdeutscher Verlag für Hochschulschriften Aktiengesellschaft & Co. KG
Dudweiler Landstr. 99, 66123 Saarbrücken, Germany
Phone +49 681 37 20 271-1, Fax +49 681 37 20 271-0, Email: info@svh-verlag.de

Copyright © 2009 by the author and Südwestdeutscher Verlag für Hochschulschriften Aktiengesellschaft & Co. KG and licensors
All rights reserved. Saarbrücken 2009

Printed in the U.S.A.
Printed in the U.K. by (see last page)
ISBN: 978-3-8381-0663-2

# Contents

| | | |
|---|---|---|
| | Abbreviations | VII |
| 1 | Motivation | 1 |
| 2 | Introduction | 3 |
| | 2.1 Insect flight control | 3 |
| |     2.1.1 Visual flight behavior | 3 |
| |     2.1.2 Mechanosensory feedback | 5 |
| |     2.1.3 Insect flight aerodynamics | 5 |
| |     2.1.4 Systems approach to flight control | 7 |
| 3 | Approach | 9 |
| |     3.0.5 Tethered vs. free flight | 10 |
| | 3.1 Choice of model organism | 12 |
| | 3.2 The importance of time-continuous, real-time, information | 13 |
| 4 | Developed technologies | 15 |
| | 4.1 MEMS micro force sensors | 15 |
| |     4.1.1 Design requirements | 16 |
| |     4.1.2 State-of-the-art | 18 |
| |     4.1.3 Sensor design & fabrication | 20 |
| |     4.1.4 Sensor integration and readout | 22 |
| |     4.1.5 Sensor calibration | 23 |
| |     4.1.6 Experimental implementation: fly tethering | 26 |
| |     4.1.7 Data analysis | 27 |
| | 4.2 Digital wing beat analyzer | 28 |
| |     4.2.1 Approach | 28 |
| |     4.2.2 High speed vision overview | 29 |

|  |  | 4.2.3 | Existing solutions . . . . . . . . . . . . . . . . . . . . . . . | 30 |
|---|---|---|---|---|

        4.2.3   Existing solutions . . . . . . . . . . . . . . . . . . . . . . .  30
        4.2.4   Design requirements  . . . . . . . . . . . . . . . . . . . . .  31
        4.2.5   Implementation . . . . . . . . . . . . . . . . . . . . . . . .  33
        4.2.6   Calibration . . . . . . . . . . . . . . . . . . . . . . . . . .  34
        4.2.7   Real-time data acquisition . . . . . . . . . . . . . . . . . .  36
        4.2.8   Results . . . . . . . . . . . . . . . . . . . . . . . . . . . .  42
        4.2.9   Discussion . . . . . . . . . . . . . . . . . . . . . . . . . .  45
        4.2.10  Concept generalization . . . . . . . . . . . . . . . . . . .  47
   4.3  Visual flight simulator . . . . . . . . . . . . . . . . . . . . . . . . .  48
        4.3.1   Other visual flight simulators . . . . . . . . . . . . . . . .  48
        4.3.2   Design requirements  . . . . . . . . . . . . . . . . . . . . .  49
        4.3.3   Implementation . . . . . . . . . . . . . . . . . . . . . . . .  49

**5 Reverse-engineering biological flight**  55
   5.1  Biomechanics of tethered flight . . . . . . . . . . . . . . . . . . . .  56
        5.1.1   Results . . . . . . . . . . . . . . . . . . . . . . . . . . . .  57
   5.2  Lift control . . . . . . . . . . . . . . . . . . . . . . . . . . . . . . .  59
        5.2.1   Methods . . . . . . . . . . . . . . . . . . . . . . . . . . .  60
        5.2.2   Experiments . . . . . . . . . . . . . . . . . . . . . . . . .  62
        5.2.3   Results . . . . . . . . . . . . . . . . . . . . . . . . . . . .  63
        5.2.4   Discussion . . . . . . . . . . . . . . . . . . . . . . . . . .  65
        5.2.5   Conclusion . . . . . . . . . . . . . . . . . . . . . . . . . .  68
   5.3  Thrust control . . . . . . . . . . . . . . . . . . . . . . . . . . . . .  68
        5.3.1   Methods . . . . . . . . . . . . . . . . . . . . . . . . . . .  69
        5.3.2   Results . . . . . . . . . . . . . . . . . . . . . . . . . . . .  69
        5.3.3   Discussion and conclusion . . . . . . . . . . . . . . . . . .  70

**6 Closing remarks**  73
   6.1  Contributions . . . . . . . . . . . . . . . . . . . . . . . . . . . . .  74
   6.2  Outlook . . . . . . . . . . . . . . . . . . . . . . . . . . . . . . . . .  75

**A MEMS calibration**  77
   A.1  Static calibration of three-axis sensors . . . . . . . . . . . . . . . .  77
        A.1.1   Calibration procedure  . . . . . . . . . . . . . . . . . . . .  77
        A.1.2   Extraction of calibration matrix . . . . . . . . . . . . . . .  77
   A.2  Dynamic considerations in displacement/force sensors . . . . . . .  79
        A.2.1   Introduction . . . . . . . . . . . . . . . . . . . . . . . . .  79
        A.2.2   Theoretical Modeling . . . . . . . . . . . . . . . . . . . . .  82

|  |  |  |
|---|---|---|
| | A.2.3 Experimental identification | 85 |
| | A.2.4 Dynamic calibration implementation | 91 |
| | A.2.5 Validation of the dynamic calibration | 94 |
| | A.2.6 Results | 94 |
| | A.2.7 As of when is a dynamic calibration necessary? | 94 |
| **B** | **QSS model of lift generation** | **97** |
| | B.1 Model variables and constants | 97 |
| | B.2 Inertial forces | 98 |
| | B.3 Aerodynamic forces | 98 |
| **C** | **Exploration of bio/robot coupling** | **101** |
| | C.1 The Cyborg system | 103 |
| |     C.1.1 Fruit fly | 103 |
| |     C.1.2 Digital wing beat analyzer | 103 |
| |     C.1.3 Fly-to-robot transfer function | 105 |
| |     C.1.4 Robot | 106 |
| |     C.1.5 Robot-to-fly transfer function | 106 |
| |     C.1.6 LED visual flight simulator | 106 |
| |     C.1.7 Overall system control | 107 |
| | C.2 Experiments | 108 |
| |     C.2.1 Naturalistic feedback | 108 |
| |     C.2.2 Amplified naturalistic feedback | 109 |
| |     C.2.3 Inverted response feedback | 109 |
| |     C.2.4 Decoupled response feedback | 109 |
| | C.3 Discussion & conclusion | 110 |
| | **References** | **113** |

# Abstract

In this thesis, I develop novel robotic tools and apply them to reverse-engineer various aspects of biomechanics and flight control in insect flight. The robotic tools are not only *novel technologies* that provide access to previously unmeasurable data, but also *analytical tools* that provide a new way to interpret the data.

The model organism I studied was the fruit fly, *Drosophila melanogaster*. Fruit flies achieve an awe-inspiring control of their inherently unstable flight despite limited neural resources. They therefore represent an important model for the understanding of sensorimotor pathways and have attracted the attention of engineers developing micro robotic devices at similar Reynolds numbers.

From a technological perspective, the spatial and temporal scales of flight control in *Drosophila* are challenging and pushed the technological limits farther. Product-level tools had to be developed to meet the demands of the biological application.

Concretely, three new tools were developed: MEMS micro force sensors, high speed vision systems, and visual flight arenas. The MEMS sensors provide a leap in the resolution ($< 1$ µN), bandwidth ($> 1$ kHz) and automation of biological force measurements.

The high speed vision implementation provides a new tool to extract wing kinematics in real time at sampling rates above 6 kHz, which is several times faster than common vision-based tracking devices.

Finally, the flight arenas were developed to stimulate the fly with precise visual patterns at high sampling rates. Together, these tools form a highly automated biorobotic platform that allows the exploration of open and closed-loop paradigms in flight control research. The tools have also been used in several complementary projects, proving their general applicability.

To apply these tools and study the sensorimotor pathways of the fly, I employed a system's level approach. This "black box" strategy requires both a precise control of the inputs and detailed knowledge of the outputs of the analyzed system. Tethered approaches, where the fly is mechanically held in place, ideally provide these desired conditions because the sensory input is well defined and the response can be measured in greater detail than in free flight approaches.

The approach was applied to four main studies of flight control: 1) A well-known biomechanical model was experimentally validated for the first time by comparing it with instantaneous flight forces. 2) I analyzed the frequency response of lift control by applying system identification tools. 3) I analyzed speed control by concurrently measuring the pitch, thrust and lift responses and compared them to free-flight measurements 4) I dynamically coupled a mobile robot with the behavior of the fly and explored the effects of various coupling strategies.

These applications provide an overall advancement in the understanding of the biophysics and neural control mechanisms of biological flapping flight. Furthermore, the developed technologies have expanded the measurement possibilities at low Reynolds numbers and high temporal frequencies.

# Abbreviations

| | |
|---|---|
| AFM | Atomic Force Microscope |
| DAQ | Data Acquisition |
| DOF | Degree Of Freedom |
| DWBA | Digital Wing Beat Analyzer |
| DOF | Degree Of Freedom |
| DWBA | Digital Wing Beat Analyzer |
| EKF | Extended Kalman Filter |
| FOV | Field of View |
| I2C | Inter-Integrated Circuit |
| INI | Institute of Neuroinformatics, ETH Zürich |
| IRIS | Institute of Robotics and Intelligent Systems, ETH Zürich |
| LED | Light Emitting Diode |
| MAV | Micro Aerial Vehicle |
| MEMS | Micro-electro-mechanical systems |
| PCB | Printed Circuit Board |
| SOI | Silicon on Insulator |
| TWI | Two Wire Interface |

# Chapter 1

# Motivation

This thesis lies in the exciting interaction between biology and robotics: novel microrobotic tools are developed and applied in a reverse-engineering approach to better understand micro aerial flight in biology.

With the advent of microrobotics in the past 20 years, novel tools are allowing to explore spatial and temporal scales that were completely inaccessible before.Engineering is also providing a set of analytical tools that can be applied to quantitatively characterize biological responses. By analyzing the biological organism with the same techniques employed for man-made devices, such approaches facilitate the transfer back to engineering [Abbott, 2007]. These "systems biology" approaches have gained a lot of momentum in recent years.

Another field that has begun to emerge recently is the one of robot-enabled biology, where robot "copies" of biological organisms have been used to test hypothesis about the biological entity, in cases where it is unpractical or even impossible to directly test it on the organism [Webb, 2006, Datteri and Tamburrini, 2007].

While engineering can improve biology, the converse is also true. As engineers strive to build smaller, more autonomous micro robots, they are faced with a series of challenges. The first one to come to mind is the miniaturization effort. Sensors, actuators, power supplies and controllers must built smaller. However, even with an exact scaled copy of a successful 'macro' device, the device is unlikely to operate succesfully, because the various interaction forces have scaled differently [Purcell, 1977]. In recent years, engineers designing micro-flying and micro-swimming robots have been slowly exiting the high Reynolds number regime, where inertial effects dominate the interaction forces, and have entered the world of intermediate Reynolds regime, where both inertial and viscous forces must be taken into account. The physical interaction with the environment becomes highly non-intuitive. As often when faced with a new challenge, engineers have turned to nature and seek for design inspirations from the solutions that were evolved trough millions of years of natural selection [1].

Another challenge linked to miniaturization lies in the control issues that arise: smaller devices are typically less stable than their macro counterparts, because the inertial effects that play an important role in passive stability have scaled unfavorably compared to other phenomena. At the same time, computing power has also shrunk because the available payload has diminished. Through scaling, the problem is therefore two fold harder: the controller has less computational power and the plant has become less stable [Rafal Zbikowski and Knowles, 2006]. Therefore, a simple yet efficient controller must be found to assure stability. Nature has evolved efficient sensory motor pathways that can generate stable and maneuverable flight despite limited neural resources. The organism studied in this thesis is a prime example of such performance.

In summary, an exciting era has currently begun: robotics and MEMS technologies are at a stage that is often compared to the IC era in the late 1970s, where they are enjoying an exponential growth and are only starting to realize their true potential [Gates, 2007]. Biology is one of the main beneficiaries of this in terms of applications. At the same time, interest in Life sciences has also recently boomed, driven by the huge potential underlying biotechnologies. This thesis is at this thrilling intersection, where robotics and biology mutually benefit each other.

---

[1] albeit all the hype of "biomimetics", there is no fundamental difference between "biomimetics" and more traditional engineering. Recently, many engineers even prefer "bioinspired" approaches, where nature's solution is used as a starting point to traditional design. I will discuss Bioinspiration and biomimetics in more detail in Section C.

# Chapter 2

# Introduction

The thesis is structured as followed. The introduction (Chapter2) will start by looking at the main biological thematic, insect flight control. It will concentrate on the main research questions addressed in this thesis. The approach (Chapter3) will describe my strategy and how it compares to other strategies.

The rest of the dissertation is divided into two main parts; a first chapter where the development of the micro technologies are described in the context of the biological application (Chapter4), and a second chapter where the technologies are applied to reverse-engineer flight control (Chapter5).

## 2.1 Insect flight control

Insects possess a highly developed flight control system that allows them to perform stabile yet maneuverable flight despite limited neural resources. This flight control system results from a highly interlinked process involving the sensory systems, the neural pathways, the motor controls and the biomechanics of flight [Dickinson, 2006].

A review of all the research done on each of these subsystems lies outside of the scope of this chapter, so I will concentrate on the more general aspects of flight control, underlining the seminal results, and point to review papers whenever possible. I will start with visual flight behavior, as it one of the most important modalities of flight control. It was also historically the first to be studied in detail.

### 2.1.1 Visual flight behavior

Vision is an essential sensor for flight. There have been entire books devoted to insect vision, so I will only present the most seminal papers that were related to our research. For a more complete overview of insect vision, see [Nakayama, 1985, Wehner, 1981]. For an overview of the relationship between vision and behavior, see [Borst and Egelhaaf, 1989, Buchner, 1984, Collett et al., 1993].

Historically, early research on flight behavior focused on vision. Kennedy was one of the first to pioneer accurate, objective quantification of animals' behavioral responses to visual stimuli [Kennedy, 1940]. Among other, he developed two important tools that are still in use today: a tethered experiment stage and a free flight wind tunnel.

The tethered experiment consisted in a tether that could freely rotate. Kennedy employed this to analyze basic optomotor responses in mosquitoes, where the insect was shown a simple visual stimulus and its body orientation was recorded. This allowed him to infer the preferred direction of flight.

Kennedy complemented these experiments with a free-flight wind tunnel, where he discovered important strategies about how visual and wind flow queues are integrated in the mosquito. These precursor technologies are remarkably similar to modern day tethered setups [Bender and Dickinson, 2006b] and wind tunnels [Fry et al., 2008].

Based on a similar systems approach, [Hassenstein and Reichardt, 1956] made important discoveries on the elementary motion detectors (EMD) of vision in tethered beetles: the famous *Reichardt-Hassenstein correlator*. Similar results were found in other insect species, such as the flying Musca [Eckert, 1973], flying Calliphora [Zaagman et al., 1977], flying apis [Kunze, 1961], flying Drosophila: [Götz, 1964], which have made this model one of the most influential in insect vision. Interestingly, this model has always been tested on tethered insects because it was technologically challenging to test it on free-flying animals. Fry recently developed a free-flight setup that allowed to test the correlator [Fry et al., 2008].

Researchers got interested in characterizing how the biomechanical performance was affected by the visual input. In his influential paper, Goetz disclosed his helicopter model, where he showed that the fruit fly, *Drosophila*, could only control the magnitude but not the direction of its average flight force with respect to its body [Götz, 1964, Gotz, 1968]. In his experiments, flies could control the azimuth of a vertical stripe by varying their torque. He showed that they elicited a strong stripe-fixation behavior, and he derived a first neural model of it.

The optomotor response and the neural circuits underlying it were further refined. Heisenberg did important experiments analyzing the role of reafference in the fly under closed-loop conditions [Heisenberg and Wolf, 1988]. Electrophysiology on the large visual neurons, the lobula plate tangential cells, provided a better understanding of the visual processing [Joesch et al., 2008]. Electrophysiology was also performed on the motor output of flies and revealed the use of small control muscles, driven typically once per wing beat, to control the large power muscles in the thorax [Heide and Gotz, 1996, Balint and Dickinson, 2001].

### 2.1.2 Mechanosensory feedback

There are numerous mechanosensors along the fly body. They fullfill different roles. Most notably, the hind wings of the fly, the halteres, lost their aerodynamic function and are employed as gyroscopes [Nalbach G, 1986].

[Dickinson, 1999] showed that the halteres were responsible for much of the low-level stabilization. Visual feedback therefore had to overcome the stabilizing effect of the halteres.

Since this this is not specifically focused on mechanosensors, please refer to [Fayyazuddin and Dickinson, 1996] for more information.

### 2.1.3 Insect flight aerodynamics

Humans have always been fascinated by the ability of insects to produce enough lift to stay aloft. A flying insect and its reciprocating wings constitute a complex physical system, intricately linking wing motion, forces and behavior. Early back-of-the-enveloppe estimations led to the expression still popularly employed nowadays that "science proved that bumblebees cannot fly", although no scientific paper was directly published on this (see review by [Weisfogh and Jensen, 1956]). Researchers started using theory originally developed for airplanes and applied it to the study of insect dynamics [Jensen, 1956]. An important assumption inherent to this work was that the flapping wing could be decomposed as a series of instantaneous steady-state conditions. This assumption is called the "quasi-steady assumption" [1].

The quasi-steady assumption was controversial, because it could not be tested experimentally at the time: all force measurements on live animals were made by averaging the forces over several wingbeats e.g. [Götz, 1964]. The measurement of instantaneous forces first appeared at the end of the 1970s in larger insect species, such as the locusts [Cloupeau et al., 1979], or blow flies [Buckholz, 1981]. A detailed review of the available work led Ellington to oppose the "quasi-steady" assumption made by Weis-Fogh [Ellington, 1984a]. In his influential series of papers, Ellington developed a comprehensive analytic theory based on the unsteady vortices [Ellington,

---

[1] Weis-Fogh did include one "unsteady" effect in his models: the clap and fling effect where the wings generate extra lift through the physical interaction at the end of the upstroke.

1984c]. The model, however, could not estimate the instantaneous forces on airfoils and was used to characterize average forces.

Instantaneous force measurements on insect remained sparse. In fruit flies, the first instantaneous measurements were published by [Zanker and Gotz, 1990], who used a taut wire to record the vibrations induced by the forces. However, these first measurements were of limited use because they were uncalibrated (forces given in relative units) and the dynamic effects of the wire's vibration were not taken into account.

Dickinson employed a similar taut-wire system to measure lift and thrust responses in fruit flies [Dickinson and Gotz, 1996]. This time, the system was calibrated and the measurements were synchronized with wake dynamic measurements to provide a general view of the aerodynamic and inertial effects taking place.

While these measurements paved the way, they also showed how difficult it was to accurately measure forces directly on an insect. These challenges led to the application of "indirect" force measurement techniques: computational fluid dynamics (CFD) and physical wing models.

These "indirect" techniques provided powerful analytic tools. Hypotheses about lift enhancing mechanisms can be tested over large parameter spaces, where it would be impractical or even impossible to test on the insect directly. For the results to be meaningful, the physical or computational models must obviously rely on precise measurements of the insect's geometric and kinematic parameters, a task that remains challenging given the small dimensions and high temporal frequencies involved [Sane, 2003]. For CFD, intermediate Reynolds number simulations are computationally more expensive than other simulations, because neither the inertial nor the viscous components of the Navier-Stokes equation can be neglected [Sun and Lan, 2004, Ramamurti and Sandberg, 2007].

The physical wing model paradigm, in particular, allowed a quantum leap in the understanding of insect aerodynamics. By dynamically scaling the wing model to a larger size, force sensors could be directly integrated at the base of the robotic wing. This is an advantage compared to tethered force measurements, where one measures the resultant force on the insect's thorax and therefore cannot isolate the indivudal body or wing components. Furthermore, by replacing the wings with equivalent masses, one could further isolate the aerodynamic from the inertial components [Dickinson et al., 1999]. These measurements led to the formulation of the so-called "Revised quasi-steady" model of insect flight [Sane and Dickinson, 2001, Sane and Dickinson, 2002], that reintroduced quasi-steady assumptions as the leading model of lift generation.

## 2.1. INSECT FLIGHT CONTROL

The dynamically-scaled wing was also used to replay kinematics measured in free flight, which showed that viscous effects during the rapid turning maneuvers of flies was more important than previously thought [Fry et al., 2003]. By comparing the forces generated in free flight to those generated in tethered flight, a quantitative comparison was made of the disrupting effects of tethering [Fry et al., 2005]. Fry showed that these disruptions were stereotyped and repeatable. In other words, tethered force measurements do not represent free flight forces, but a direct quantitative mapping exists between the two.

Although the aerodynamics of insect flight has largely been understood with the help of the dynamically scaled measurements, a validation of the "revised quasi-steady" model on direct force measurements was never done, most likely because of the difficulty associated with simultaneously measuring *instantaneous* flight forces and kinematics. Such a validation is undertaken in this thesis, by using the novel tools we developed (see Section 5.1). For a more detailed review of insect flight aerodynamics, see [Sane, 2003].

### 2.1.4 Systems approach to flight control

The fly is becoming more and more a model organism to explore systems-level processing [Dickinson, 2006]. This approach was motivated by the increasing knowledge that the subsystems constituting the fly's sensorymotor pathways were highly interlinked [Dickinson et al., 2000]. It became less relevant to study each subsystem individually. Instead, the organism was studied as a black-box controller [Frye and Dickinson, 2004].

Aeronautical engineers started to get interested in the problem. Taylor applied state-space modeling to characterize the stability of tethered locusts using stroke-averaged forces [Taylor and Thomas, 2003]. In a subsequent paper, he performed the same analysis with instantaneous forces [Taylor and Zbikowski, 2005]. Zbikowski has been applying reverse-engineering approaches to the study of insect flight control [Rafal Zbikowski and Knowles, 2006]. Tanaka used a rigorous approach from aeronautics to characterize the lift controller in bumblebees [Tanaka and Kawachi, 2006].

A key aspect in these systems approaches relies in the ability to precisely define the inputs and accurately measure the outputs [Taylor et al., 2008]. Given the speed at which control commands occur [Balint and Dickinson, 2004], stroke averaged measurements generally do not provide sufficient detail. Furthermore, while open-loop techniques provide a standard method to identify systems, closed-loop verification is often necessary to validate the identified system. These two characteristics require time-continuous and real-time measurement techniques, which will be an integral part of the technology development in this thesis.

# Chapter 3
# Approach

To address some of the relevant research questions presented in the previous chapter, a decision had to be made on the type of experiment to be implemented. A key choice is whether the experiments should use tethered or free-flying insects.

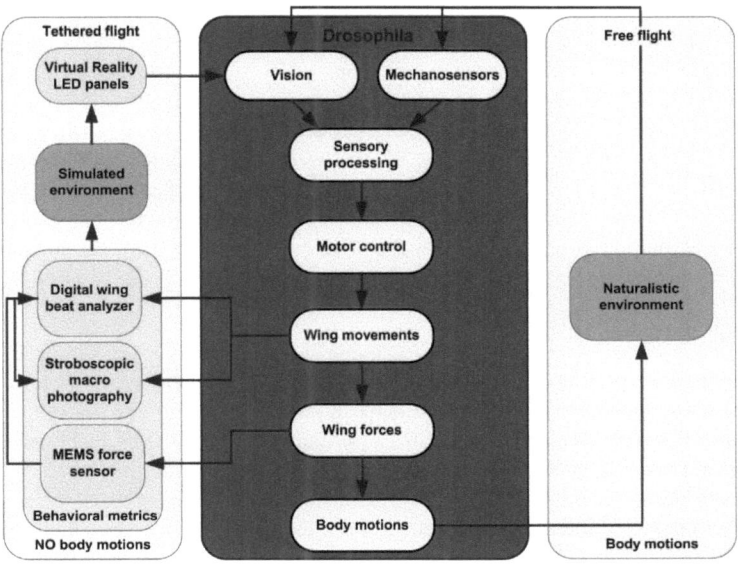

**Fig. 3.1:** An overview of the research framework. See the text for details.

Taking a somewhat reductionist view, we can consider the fruit fly as black-box encompassing sensory systems, neural pathways and motor output (see Fig.3.1) In free flight conditions, the fly interacts with the environment and receives natural feedback via its sensory systems (Fig.3.1, right). Free flight is intrinsically closed-loop, and open-loop conditions must be artificially created [Fry et al., 2008]. Tethered flight is intrinsically open-loop, and closed-loop conditions must be simulated (see Fig.3.1, left, and [Heisenberg and Wolf, 1988]).

### 3.0.5 Tethered vs. free flight

Tethered and free flight approaches each have pros and cons. A nice thing is that these largely complement each other. Ideally, one should therefore verify results from tethered experiments in free-flight and vice-versa.

In general, tethered approaches provide more detailed knowledge of the sensori-motor processes. The fact that the fly is attached allows to precisely control what its sensory inputs are, and to measure in great detail what how the fly is reacting. Tethered flight provides ideal open-loop conditions.

Free flight provides naturalistic conditions and is therefore less prone to artifacts. Free flight is ideal for closed-loop experiments.

In my work, I worked with a tethered setup. This was largely motivated by the fact that we had a new type of sensor that could measure tethered flight forces. However, the tethered approach also allowed us to measure low-level flight control aspects that would have been impossible to measure in free flight: kinematics and dynamics of insect flight are notoriously difficult to measure due to the high beating frequency and low magnitude of changes. The tethered setup therefore provided an ideal method to explore flight control using a systems approach (See Section 2.1.4).

**Tether artifacts**

Tethering disrupts the natural feedback of flight. It is therefore essential to understand what these effects are and how one can cope with them. The insect's various sensors all receive some form of feedback during flight. In tethered flight, however, the visual system, the antennae and especially the halteres do not receive the normal feedback during maneuvers. There have been workarounds implemented in the past, that all involve artificially closing the feedback loop. For vision, closed-loop turning response experiments have been made for many years, where the fly's yaw torque is used to drive the azimuth position of a stripe [Götz, 1964]. For the antennae and halteres, flies have been tethered in a gimbal device that rotates the tethered fly up to rotational speeds of 2000°/s [Sherman and Dickinson, 2003]. There have recently been experiments aimed at creating artificial Coriolis forces on the halteres by actuating the tether with force patterns that are phase-locked with the wing beat [Bartussek et al., prep]. Flies have also been tethered to a needle with a special bearing that allowed it to freely move around the yaw axis [Bender and Dickinson, 2006b, Bender and Dickinson, 2006a]. These methods are powerful, but do not completely replace a free-flight verification because they always rely on a model of the behavior. The closed-loop turning response experiments, for instance, have been controversial: how should the turning torque affect the position of the fly? The biomechanics underlying this question are not so trivial, because a fly is beating its wings as it turns. In such artificial closed-loop experiments, there is therefore always the risk that one of the meaningful feedbacks is not provided correctly. [Taylor et al., 2008] argues that tethered experiments are most meaningful in open-loop [1], where the feedback is missing by design, but where it is acknowledged to be so.

Another effect of the tether is that it alters the mechanical properties of the thorax. The thorax resonates at the wing beat frequency, driven by the large power muscles. This resonance is conveyed further to the wings via the control muscles acting on the wing hinge structure. The tether glued to the thorax obviously affects the biomechanical properties of this system. One would expect the rigidity of the structure to go up, which would result in a higher resonant frequency. This fact is supported by experimental evidence: tethered fruit flies often have wing beat frequencies around 230 Hz instead of about 200 Hz in free flight.

Finally, the tether imposes a fixed position to the body. During the course of a wingbeat in free flight, the body may move slightly to generate some inertial forces or change the center of gravity. These movements are not possible in tethered flight. If the fly is tethered to a device with a movable structure, such as a force sensor based on spring elements, then the dynamics of the movable structure add to the normal body dynamics. The tether also imposes a fixed orientation with respect to gravity. It is unclear how important of a role gravitotaxis plays in flies, but the difference between the fixed orientation and the preferred orientation with respect to gravity may result in a constant pitching torque in the fly.

A recent study obtained a good characterization of tethering effects by visually comparing wing stroke patterns in free and tethered flight in otherwise identical

---

[1] Taylor does not use the term "open-loop" and prefers to use the expression "broken-loop" to emphasize that the fly is still treating its sensor information as feedback. I will stick with "open-loop" as it is the most widely used in the literature. This is purely a nomenclature definition and does not affect the interpretation of the results.

experimental conditions [Fry et al., 2005].

In summary, the artifacts created by tethered conditions must be carefully considered. The choice of experiment is crucial to avoid misinterpretations. Free flight verification is highly desirable. Obviously, it is usually not possible to exactly replicate the results in free flight (there would have been no need for a tethered approach), but the results can normally be generalized to a level where comparison is possible.

## 3.1 Choice of model organism

We decided to work with the fruit fly, *Drosophila melanogaster*. The fruit fly has been a model organism in biology for over a century. It was originally popular because of the ease with which it could be grown and bread [Morgan, 1915]. Fruit flies are small enough so that several colonies can be elevated in parallel in a small room; their gestation period is just around ten days, which means that one can rapidly go through several generations. Multiple phenotypes are available to distinguish breeds. The fruit fly has therefore extensively been employed for genetic studies. The resulting genetic tools can benefit other areas of biology as well, where sensorimotor processes can easily be genetically isolated. This makes the fruit fly one of the most studied organism in science, as shown by the willingness of leading scientific journals to consecrate entire issues on the organism [Gunter et al., 2007].

For flight control, fruit flies represent an ideal model system: they show robust and highly stereotyped reflexive behaviors. Their neural control is essential to maintain stability and is therefore highly specialized.

For robotics, fruit flies represent an interesting, but still distant, model for autonomous devices. Current micro-aerial vehicles operate at Reynolds numbers that are at least an order of magnitude larger [Wood, 2008]. Nonetheless, certain aspects of fruit fly flight, such as specific lift generation mechanisms, have been imitated in flapping micro aerial vehicles. Micro-swimmers may also benefit from a better understanding of the biological processes at play, as their dynamic regime overlaps that of insect flight [Abbott et al., 2007].

The control techniques employed by biological organisms are especially interesting, because they use limited neural resources to produce stabile yet highly maneuverable flight patterns [Rafal Zbikowski and Knowles, 2006]. Such efficient control techniques are very relevant to MAVs, where the small payload limits the complexity of onboard calculations.

## 3.2 The importance of time-continuous, real-time, information

In the past, insect flight studies focused mostly on large time domains, since they are much easier to access experimentally. The flight forces and wing kinematics were averaged out over several wing beats.

The neural control mechanisms, however, act within a stroke. The neural signal phase-activates flight control muscles that change how the strains in the thorax are conveyed to the wings. A lot of relevant information is discarded by only studying the average effect of these signals. Furthermore, an instantaneous analysis of the wing kinematics and forces is necessary to study the important question of how the wing movements generate the intended forces.

In this thesis, I therefore decided to develop time-continuous, real-time measurement techniques. *Time-continuous* refers to a sufficient temporal resolution to measure individual motor control commands. *Real-time* refers to the ability to extract behavioral information in real-time so that it can be used to drive external equipment or create closed-loop conditions. The real-time capability also allows to measure over extended periods of time, since only the extracted features are stored. This opens up the possibility to explore large parameter spaces in an automated fashion.

These two characteristics define key requirements for the technologies that were developed during this thesis (see Fig.3.1).

# Chapter 4

# Developed technologies

Analyzing micro aerial flight of an object the size of *Drosophila* is a challenging task. As dimensions scale down, accessibility becomes difficult. Wing beat frequency increases [Woods et al., 2001], putting temporal constraints on the measurement tools. Due to the increasing flight instability, the changes in kinematics required to generate rapid maneuvers are very subtle. Aerodynamic forces range in the micro Newtons. Furthermore, the sensory system has evolved to sample at high rates, making artificial sensory stimulation more difficult.

To overcome these challenges, we developed a set of novel microtechnologies that cope with these requirements (see Fig.3.1 for an overview). The first consisted in MEMS force sensors, which were designed to measure the instantaneous flight forces generated within a wing stroke. The second was a custom-designed high speed vision, which was designed to measure the subtle changes in the temporal and spatial properties of the wing kinematics. The third was a visual flight simulator. It was built to take into account the near-panoramic field of view of flies and the ca. 150 Hz flicker rate of vision in fruit flies.

In the following sections, I will systematically discuss, for each of these three technologies, 1) the design requirements, 2) the current state-of-the-art, and 3) the implementation of the chosen design.

## 4.1 MEMS micro force sensors

Micro-electro-mechanical systems (MEMS) were pioneered in the 1980s, when engineers took micro fabrication techniques developed for the IC industry and applied them to the design of mechanical parts. MEMS technologies have since enjoyed a formidable growth, leading to industrial applications such as the accelerometers featured in the airbag systems of cars and opening up new research avenues by allowing scientists to probe the micro world. Biology has benefited a lot of the new tools that MEMS has to offer. This thesis is one example of these applications.

### 4.1.1 Design requirements

The goal was to measure the instantaneous aerodynamic and inertial forces produced by a tethered fruit fly. Fruit flies weigh about 1 mg, and must therefore generate an average lift force of 10 µN to stay aloft. Since we are not only interested in the forces' mean value, but also in how the forces vary as a function of what the fly is seeing, we must set temporal and force resolutions limits.

**Temporal and force resolutions**

Based on previous measurements of lift forces in fruit flies [Dickinson and Gotz, 1996], the forces range from about -30 µN to +60 µN. The required resolution is more difficult to estimate: no one has accurately measured the minimal magnitude of voluntary steering force commands. We decided that a sensor capable of measuring changes of at least 10% of the body weight should be sufficient to capture most significant responses of the fly. This estimations sets the minimum resolution to around 1 µN.

Regarding the temporal aspect, the forces generated by a flapping wing vary a lot during a single wingstroke: Both ends of the stroke, pronation and supination, cause sudden peaks in the aerodynamic forces. The acceleration and deceleration of the wings add inertial loads. On top of this, control muscles act at precise wing beat phases to alter the way in which the energy from the large power muscles is conveyed to the wings. To measure these effects, the bandwidth of the force measurement device must be well above the 200 Hz wing beat frequency. Measurements by [Dickinson and Gotz, 1996] showed that significant information was still present at frequencies as high as 4 kHz.

**Relevant flight force/torque components**

Given the fly's ability to maneuver in all six directions of translation and rotation, it is useful to review which are the most relevant forces and torques to flight control. Historically, yaw torque has been the most measured component, because it corresponds to the fly's intention to turn and therefore represents the most robust behavioral response. However, yaw turning in tethered flies is very different than yaw turning in free-flight, because the feedback to the halteres has been disrupted. Although a few groups have artificially recreated this feedback, either by rotating the entire tethered setup [Sherman and Dickinson, 2003], or by generating pseudo-Coriolis forces to the fly [Bartussek et al., prep], these techniques are cumbersome to use with a concurrent measurement of forces because they generate an additional force that must be separated from the insect's forces. For these reasons, I decided not to work on yaw forces and concentrated on other components of flight.

Lift force is the largest component of flight forces. The fly must continuously adjust lift throughout maneuvers to compensate for gravity. Much attention has been focused on lift in fruit flies: from stroke-averaged tethered measurements [Gotz, 1968], to calibrated instantaneous forces [Dickinson and Gotz, 1996], to free-flight force estimations through robotic wing models [Fry et al., 2003]. In this thesis, I will continue to look at lift, but I will put more focus on the control of lift as a response to visual inputs. Previous measurements were all limited to a few experiments because single measurements were cumbersome to obtain. The automated setup that I developed in the course of this thesis allows to explore a much larger parameter space.

Thrust forces are important to understand the sensorimotor pathways involved in forward flight. Goetz studied stroke-averaged thrust components and put it in relationship with the lift forces in his influential "helicopter model" [Gotz, 1968], while Dickinson measured the instantaneous components of thrust forces in *Drosophila* [Dickinson and Gotz, 1996]. The helicopter model implies that the mean flight force vector always has the same orientation with respect to the body. In other words, forward speed must be initiated first by a pitch angle adjustement, followed by an increase in overall force production. This makes a tethered analysis of forward velocity challenging, because the fly cannot change pitch angle, so the measured thrust response might just be an artifact of the fly trying to pitch forward (and not making it) and might have very little in common with a natural thrust response.

As just mentioned, pitch torque is a central component forward velocity control. [Gotz, 1968] measured average pitch forces, but no measurement of instantaneous pitch forces have been done to our knowledge.

The two remaining components of flight, side slip force and roll torque, have a more minor role in insect flight. They are typically not associated with maneuvers, and are therefore controlled mainly to adjust stability along those axis. I have not concentrated on either of these components of flight.

## 4.1.2 State-of-the-art

Here I will discuss the suitability of various force sensing techniques to the problem at hand. I will also review previous force measurements made on *Drosophila*.

**General classes of force sensing**

Various micro electromechanical systems (MEMS) force sensor have been realized:

- **Piezoelectric** [Rao et al., 1995]: Piezoelectric force sensors have micro-Newton resolutions. Their main advantage is that they are robust with respect to handling.

- **Piezoresistive** [Fahlbusch et al., 2002]: Piezoresistive force sensors have been widely used in micro force sensing and gripping. [Rangelow et al., 2002] have built 3DOF micro force sensors with a resolution of 1 nN.

- **Capacitive** [Sun et al., 2003]: 10 nN force resolutions have been reported. Compared to the other methods, capacitive microforce sensing has the advantage of low power requirements, no hysteresis and low sensitivity to temperature variation.

- **Magnetic** [Gibbs et al., 2004]: Force sensors based on magnetostriction have been developed, but to our knowledge, resolutions below 0.1 N have not yet been attained.

- **Scanning probe microscopy** Scanning probe microscopy, a technique to map the surface of a material, relies on the interaction force of a probe and the specimen. The most common type of scanning probe microscope, the atomic force microscope, measures this force through the deflection of a cantilever beam. Different optical methods, such as laser speckle or interference, are used to measure the deflection. Scanning probe microscopes have extremely fine force resolutions. Their main disadvantage lies in the limited accessibility of the probe, which is surrounded by the scanning actuation mechanism, and the complexity and cost of the whole device.

In view of the characteristics cited above, piezoelectric and magnetostriction are not a viable option for fruit fly force measurement because their resolutions are currently too coarse. The other three transduction modalities, piezoresistive, capacitive and scanning probe microscopy, have all been used to measure forces produced by tethered fruit flies:

## 4.1. MEMS MICRO FORCE SENSORS

**Table 4.1:** Previous reported force & torque measurements on tethered *Drosophila*

| Reference | Force component | Temporal res. | Sensor type |
|---|---|---|---|
| [Götz, 1964] | yaw | stroke averaged | "torque meter" |
| [Gotz, 1968] | lift; thrust; | stroke averaged | "force meter" |
| [Heisenberg and Wolf, 1979] | yaw | stroke averaged | closed-loop "torque meter" |
| [Zanker and Gotz, 1990] | lift | non calibrated instantaneous | taut wire |
| [Dickinson and Gotz, 1996] | lift & thrust | instantaneous | taut wire |
| [Nasir et al., 2005] | lift, thrust & pitch | no force trace | piezoresistive sensor |

**Previous applications to fruit fly force/torque measurements**

Table 4.1 gives an overview of the force/torque measurements made on fruit flies.

As we can see in this table, there have been three main strategies employed to measure forces and torques in tethered *Drosophila*:

1. The **torque meter** used the same physical principle as a galvanometer, where an electrical current creates a rotary deflection. For the torque meter, the reciprocal principle is employed: the torque generated by the tethered fly was measured through the electrical current needed to keep the fly pointed in the same direction [Götz, 1964]. The current-to-torque relationship was shown to be linear. In a subsequent paper, this same principle was adapted to measure separately lift and thrust [Gotz, 1968]. Numerous studies have used improved implementations of the Goetz apparatus to measure stroke-averaged turning torques [Heisenberg and Wolf, 1988] [Hesselberg and Lehmann, 2007].

2. The **taut-wire** device employs a similar principle to the atomic force microscope. The fruit fly is tethered to a wire under tension. The lift forces produced by the fly cause the wire to vibrate. These tiny vibrations are picked up by shining a laser on the wire and observing how the diffraction pattern moves with the help of a phototransducer [Zanker and Gotz, 1990]. The taut-wire method was somewhat cumbersome to employ since it required careful calibration of

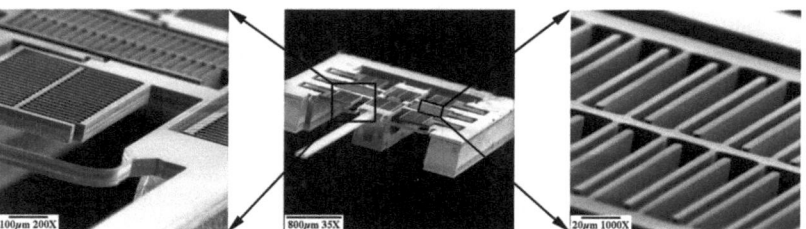

**Fig. 4.1:** SEM view of the MEMS technology employed by IRIS (taken from [Sun and Nelson, 2004])

the dynamic characteristics for each new fly [Dickinson and Gotz, 1996]. The size of the device also prohibited its integration into flight simulators.

3. **piezoresistive MEMS sensor** have been employed to measure flight forces of tethered fruit flies [Nasir et al., 2005]. The paper's results show that the resolution of the sensor was too coarse to measure the flight forces reliably. The only detectable signal was a predominant 200 Hz component in the FFT.

### 4.1.3 Sensor design & fabrication

A capacitive MEMS sensor was chosen to measure the instantaneous flight forces of fruit flies. Capacitive sensors have the advantage of high sensitivity, low drift and temperature invariance. Sun was the first to demonstrate direct measurements of fruit fly flight forces using MEMS sensors [Sun et al., 2005].

The majority of MEMS design and the totality of MEMS fabrication was done by Felix Beyeler and Simon Muntwyler at IRIS. A close collaboration was maintained to ensure that the sensors would meet the requirements for the fruit fly measurements. There were two types of sensors used for the fruit fly project: the 1 DOF, employed for lift measurements, and the 3 DOF, employed for concurrent lift, thrust and pitch measurements. The key parameters of design are the spring shapes and the inter comb distances. The sensors are fabricated from SOI wafers and go through a four step fabrication process. For more detail about sensor design and fabrication, see [Beyeler et al., 2008].

## 4.1. MEMS MICRO FORCE SENSORS

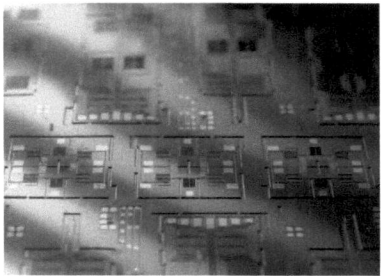

**Fig. 4.2:** Wafer with completed sensors. This image shows different 3DOF force sensors after the four micro fabrication steps have been completed.

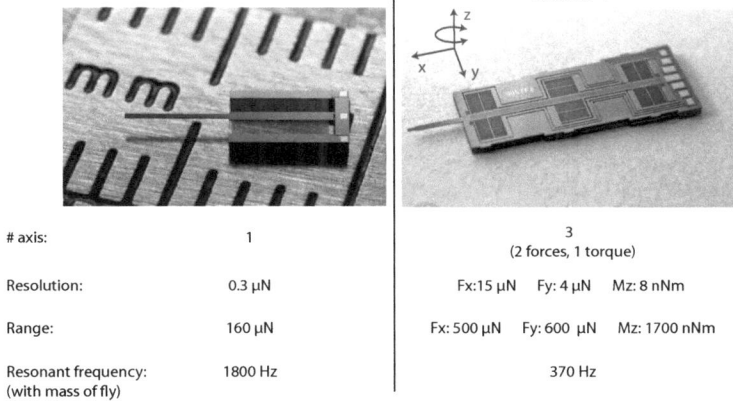

**Fig. 4.3:** Overview of the two types of MEMS sensors used to measure flight forces on the fruit fly. a) The single axis force sensor produced by the ETH startup FemtoTools (www.femtotools.com) was used to measure lift forces. b) The three axis force sensor IRIS FF1 designed and fabridcated by Simon Muntwyler during his Master thesis [Muntwyler, 2006]

### 4.1.4 Sensor integration and readout

Once fabricated, the sensors are individually removed from the wafer and glued to a PCB with epoxy glue. The sensor is electrically connected to the PCB by manually wire bonding the connection pads.

An electronic readout was designed to measure the vast varying capacitance. The readout circuitry was an important part of this project because the fruit fly application combined fine resolution with high sampling rate requirements. The readout circuitry had to sample very rapidly while keeping the electrical noise low. As opposed to static applications, it was not possible to improve the signal-to-noise ratio by low-pass filtering the data because the measured signal contained important information at high frequencies. Finally, the multi-axis measurements increased the effective sampling rate because each capacitive pair had to be measured sequentially.

Two readout circuitries were implemented. The first is ideal for single axis sensors, but is not adapted for multi-axis. The second one was designed for multi-axis.

**Single-axis readout circuit**

The first readout is based on the MS3110 chip, which converts the difference in capacitance of two capacitors into a voltage. The main advantages of the readout circuitry are that it is simple to set up and the voltage output is easy to measure. The main drawbacks are that its sampling rate is limited to 8 kHz and it is only applicable for one capacitance pair. For multi-axis sensors, a MS3110 would be required for each capacitance pair. Furthermore, the ground voltages would have to be separated for each capacitance pair, because MS3110 varies this voltage for each measurement cycle. The separation of the grounds becomes very challenging for MEMS sensors above 2 DOF, because there are only limited physical connections to the movable part of the sensor, which acts as the ground. The total number of electrical pads connecting the sensor to the PCB is $3 \cdot n$, where $n$ is the number of DOFs.

**Multi-axis readout circuit**

The second readout system measures the discharge time of each capacitor individually. The main advantages of this readout are that it is extremely fast at measuring the capacitance, and doesn't require a separation of the ground voltages. Its main drawbacks are that it is more complex, and provides results in a digital format (SPI) that has to be interpreted by some subsequent IC. This makes this readout ideal for multi-axis sensors. The number of electrical pads connecting the sensor to the PCB

## 4.1. MEMS MICRO FORCE SENSORS

**Fig. 4.4:** 1DOF sensor and integrated readout, as sold by IRIS startup Femto-Tools. The readout provides a voltage proportional to force.

**Fig. 4.5:** Multi-axis readout concept. The MEMS sensor is plugged into the multi-axis readout that uses the chip PSØ21 (see Fig.4.6). An intermediate device packages the measurements into manageable sets for a normal computer (see Fig.4.9)

is $2 \cdot n + 1$.

**Acquisition**

For both readout circuits, digital acquisition boards were required to record the data. For the single axis readout circuit, a standard DAQ card was sufficient to capture the full range and resolution of the MS3110 chip at 8 kHz. For the multi-axis, the DAQ card needed to be more specific (see Fig.4.9), since it had to communicate to the readout via a SPI interface. The SPI interface is first used to set the measurement registers on the readout chip. The readout is then ready to acquire data. Each measurement starts with the DAQ asking the readout to perform a measurement and the DAQ responding with the data.

### 4.1.5 Sensor calibration

The raw data from the readout has a relationship to a force. This relationship depends on the mechanical characteristics of the sensor, such as its spring stiffness, and on the electrical characteristics of the sensor and readout circuit, such as the size of the capacitors and the strategy employed to convert this capacitance to a signal. Because

**Fig. 4.6:** A: Multi-axis force sensors. B: multi-axis readout circuits. The force sensors plug into the readout. The readout sends out digital data to an acquisition box (see Fig.4.9).

**Fig. 4.7:** The "nutshell" was designed to gather the SPI measurements sent from the multi-axis readout and package them into UDP packets that are sent via ethernet to a host computer. It contains a commercial microcontroller board (red) with an additional application-specific PCB layer to facilitate the connections.

## 4.1. MEMS MICRO FORCE SENSORS

each sensor is slightly different, the force/signal relationship must be characterized individually.

The MEMS sensor measures forces in a indirect way. The forces applied at the probe tip deform the spring elements and cause a relative displacement between the "moveable" combs and the "rigid" combs. In turn, this displacement changes the electrical capacitance of the comb pairs, which can be sensed by the electronics mentioned above. When low frequency forces are being applied to the tip, the forces and displacements can be considered to be proportional. As the frequency of the input forces increase, the forces start exciting vibrations in the sensor. The forces and displacements are not proportional anymore. This leads to two types of calibrations, depending on the frequencies one is measuring. The first calibration procedure is the static calibration, and assumes a proportional force/displacement dependency. It is a simple calibration procedure, because it does not depend on the time course of the forces. The second calibration procedure is called the dynamic calibration. It is intrinsically more complex, since it involves sweeping through the frequencies and characterizing the dynamic response of the sensor. The following table gives an overview on the three main calibration strategies that were employed.

**Table 4.2:** Main Characteristics of the static and dynamic calibration for single-axis and multi-axis sensors. The mechanical, electronic and resulting calibration equations are presented. $F$ is the applied force on the probe. $x$ is the resulting deformation of the sensor. $k$ and $K$ represent the spring constants, $g$ the dynamical model of the force/displacement characteristic. MS3110 represents the single-axis readout circuit, PS021 the multi-axis one. $k_1$ and $K_1$ represent the static calibration gains.

|  | Static | | Dynamic |
|---|---|---|---|
| # axis | 1DOF | 3DOF | 1DOF |
| Mechanic | $F = k \cdot x$ | $\boldsymbol{F} = \boldsymbol{K} \cdot \boldsymbol{x}$ | $F = g^{-1}(x, \dot{x}, \ddot{x}, t)$ |
| Electronic | $V = \text{MS3110}(x)$ | $\boldsymbol{C} = \text{PS021}(\boldsymbol{x})$ | $V = \text{MS3110}(x)$ |
| Calibration | $F = k_1 \cdot V$ | $\boldsymbol{F} = K_1 \cdot \boldsymbol{C}$ | $\widehat{F} = f^{-1}(V, \dot{V}, \ddot{V}, t)$ |

**Static calibration**

The static calibration procedure assumes the sensor's electrical and mechanical characteristics can be reduced to a time-independent force/signal relationship. For the single axis case, the sensor transduction is furthermore linear, meaning this relationship can be completely described through a gain and an offset. To determine the gain and offset, a reference sensor is used. The reference sensor is brought in contact with the probe of the MEMS sensor, and the sensors' outputs are recorded synchronously. A line is then fit through the force/signal plot.

The multi-axis calibration is a little more complicated for several reasons. First, there are no multi-axis reference sensors, so our strategy was to use a single-axis reference sensor and repeat the calibration for multiple directions. Second, pure torques are difficult to generate, so the torque calibration must be done by applying forces at precise points. Finally, the transduction properties of a given axis are not independent of the forces/torques acting on the other axis, so that a calibration matrix is in theory necessary for all combinations of forces and torques.

Our efforts at calibrating multi-axis force/torque sensors are described in Section A.1.

**Dynamic calibration**

The dynamic calibration is more complicated than the static calibration. Ideally, a sinusoidal force source would be used, and the frequency response of the sensor characterized by varying the frequency of the sinusoid. Unfortunately, such an ideal force source does not exist, at least at the force scales we are looking at. We therefore used an alternative method to identify the mechanical and electrical models of the sensor. These models are then implemented into a set of differential equations to compute the forces from the capacitance measurements.

The model identification is explained in detail in the appendix Section A.2.

### 4.1.6 Experimental implementation: fly tethering

Once the force sensor has been mounted on the PCB and the whole system calibrated, the biological experiments could begin. Much care had to be taken to design an appropriate tethering stage, because this was the most critical process in view of the sensors' fragility. The tethering stage consisted in a spherical gimbal with three rotational degrees-of-freedom and a micromanipulator with three translational degrees-of-freedom, giving full control of the positioning (see Fig.4.8). These devices were mounted below an optical microscope.

Flies were first cooled down to 4 °C in a refrigerator. At this temperature, flies are anesthetized but remain alive and will recover within a few minutes. They were then placed in the gimbal device. The gimbal device consisted in an aluminum block with a semi-sphere dug into it. The aluminum block was cooled to 4 °C to keep the fly anesthetized. It was cooled with a Peltier thermoeletric heat pump. The heat

## 4.1. MEMS MICRO FORCE SENSORS

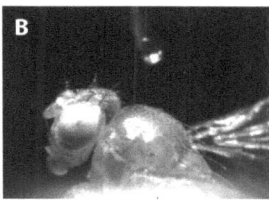

**Fig. 4.8:** Tethering setup. A: The fly is placed in the sarcophagus of an aluminium semi-sphere that allows the tethering angle to be precisely adjusted. The sphere is placed in an aluminium bloc that is cooled by a Peltier heat pump.

pumped from the aluminum block was evacuated by a fluid cooling system, originally designed for high end computer cooling. The semi-sphere contained a small slit at its center, designed to hold the fly in position. The sphere could be rotated in its casing, allowing the fly's angular position to be adjusted. This was extremely important to ensure a proper tethering angle. Furthermore, with the three-axis sensor, the fly had to be tethered from a sideway angle, because otherwise the sensor would have obstructed the vertical field-of-view of the microscope. The micromanipulator consisted in a three axis linear stage with a MEMS holder at the end.

To tether the fly, a tiny droplet of UV curable glue (Loctite Clear Glass) was placed on the fly's thorax. The tip of the MEMS sensor was then brought in contact with the glue. A UV light source was used to cure the glue. The sensor with the attached fly was lifted up and the fly was given a few minutes to recover. Flies usually initiated flight spontaneously or with a gentle air puff.

After the experiments were performed, the fly was removed from the sensor, either by gently brushing it off, or by dissolving the glue using a small amount of Acetone.

### 4.1.7 Data analysis

The measurements were saved in ASCII text format and interpreted using a set of MATLAB routines. Each measurement series was logged into a Research book and was then digitized in a "measurement selection" file, where all measurement timestamps were entered. The routines would use the measurement selection file as

28                                    CHAPTER 4. DEVELOPED TECHNOLOGIES

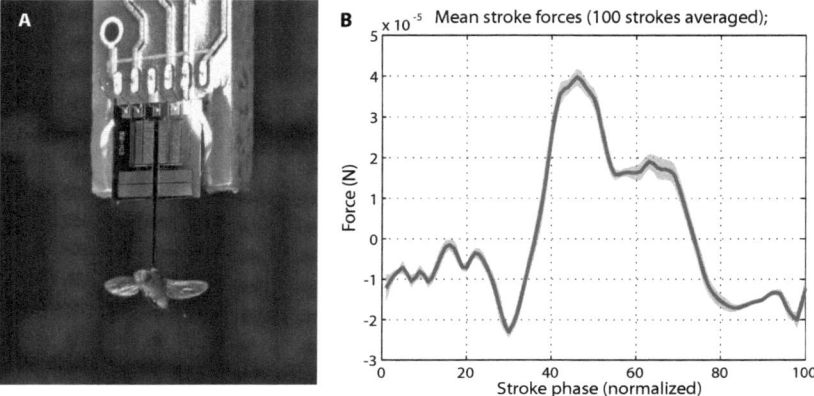

**Fig. 4.9:** A: Fruit fly tethered to force sensor. B: Instantaneous force measurements (mean ± S.D. for 100 consecutive wing strokes

list of measurements to analyze.

## 4.2  Digital wing beat analyzer

The measurement of instantaneous flight forces are of limited use if one doesn't know when they occurred within the wingstroke. A detailed analysis of the wing kinematics also provides us with a more direct view of the flight motor output of the fruit fly. A kinematic analysis is therefore the perfect complement to the MEMS force measurements. The combination of the two allows to measure and understand i) the motor output from the neuromuscular system ii) the biomechanics of flight and iii) the resulting flight maneuvers. Here, we present a novel system developed in the scope of this thesis to extract the wing kinematics in real time.

### 4.2.1  Approach

In prior efforts, the measurement of wing motion has been performed using an infrared light that projected the shadow of each wing through a crescent mask onto a photodiode receptor [Dickinson et al., 1993, Reynolds and Riley, 2002]. While this setup can measure the relative stroke amplitude of both wings, it is fundamentally limited

## 4.2. DIGITAL WING BEAT ANALYZER

by the reduction of a three degree of freedom movement, i.e. the wing angles, to a single dimensional analog value, i.e. the photodiode readout voltage. Furthermore, construction of the analogue hardware and the requirement for a precise alignment of the fly prior to measurement have hindered a more wide-spread use of this technique.

We opted for a new system based on computer vision. Computer vision has the advantage of being highly flexible, allowing additional functionality, such as automatic calibration features to be implemented. In addition, the two dimensional view combined with prior knowledge of an organism's dimensions can be used to obtain new metrics, such as a full 3D reconstruction of the wings' positions.

### 4.2.2 High speed vision overview

Many processes in technology and science require systems to be controlled in real time using a contact-free measurement based on visual techniques. Production line processes may require faulty components to be identified visually and subsequently removed. Visual servoing requires manipulator and object positions to be extracted from live images to generate appropriate control commands [Imai et al., 2004]. External vision systems are also employed to locate micro-robots [Yesin et al., 2006], which can typically not be equipped with position sensors. Real-time, contact-free measurement techniques also play an increasingly important role in biology to track moving organisms while controlling aspects of the experimental process [Ogawa et al., 2005, Fry et al., 2004].

These processes share the requirement to sample consecutive images, process them on a frame-by-frame basis and use the result of the image analysis to take appropriate action for the process control. Real-time high speed image analysis is technically demanding and underlies constraints with respect to the amount of information that can be transferred and processed between frames. The system bandwidth BW [bytes/s] is given by[1]:

$$BW = (f \cdot d \cdot FOV)/\sigma^2 < BW_{max} \tag{4.1}$$

where $f$[Hz] is the sampling frequency, $d$[bytes] the depth of each pixel, $FOV$[m$^2$] the field-of-view and $\sigma$[m] the smallest resolvable object size. The amount of data generated at each frame is dictated by the application specific requirements for the

---
[1] For these equations, we assume that the smallest detectable change of position is one pixel wide (no subpixel resolution). For simplicity, we also assume a FOV and camera sensor that have the same width and height.

spatial resolution and FOV, which are related to the vision system's pixel array size and optics as follows:

$$\begin{cases} \sigma = (z \cdot p)/F < \sigma_{max} \\ FOV = N\sigma^2 > FOV_{min} \end{cases} \quad (4.2)$$

where $F$ is the lens focal length, $z$ is the lens working distance, $p$ [m] is the pixel side length on the camera sensor and $N$ is the total number of pixels on the camera sensor.

Bandwidth limitations become particularly restrictive when tracking small objects, which tend to move at higher relative velocities and cover larger areas relative to their size than larger objects. Consequently, $f$ and $FOV/\sigma^2$ need to be increased, respectively, and the bandwidth, therefore, quickly becomes a limiting factor in practical applications.

To avoid bandwidth limitations, a general strategy of transferring only relevant information is clearly beneficial. Generally speaking, relevant information in tracking applications is often localized in space and time, and, therefore, selective image sampling and analysis should provide a powerful strategy to avoid bandwidth limitations in real-time vision tracking applications. In this paper we describe concepts and techniques applied to implement a high performance real-time high-speed vision application based on a selective image sampling and analysis strategy. This concept is implemented using standard digital vision hardware and, thus, provides a flexible and affordable solution implemented in software.

Our solution relies on a process model that provides the spatial and temporal information required to select relevant pixel locations for analysis. We only expose and transfer the relevant portion of the image using a dynamic region-of interest (ROI) that varies in both position and size, which avoids being limited by image transfer bandwidth. In quantitative terms, we can replace the *global* FOV in (4.1) with a smaller *local* FOV, allowing the BW budget gain $FOV_{global}/FOV_{local}$ to be used to achieve both a fine spatial resolution and a high frame rate, thus making efficient use of the available system bandwidth (see comparative diagram of Fig.4.10). The applied concepts are general and can be implemented in standard high speed video hardware in applications where bandwidth is a limitation.

### 4.2.3 Existing solutions

Taking inspiration from early tracking work [Weiss et al., 1987, Allen et al., 1993], various approaches have emerged that take advantage of recent advances in imag-

## 4.2. DIGITAL WING BEAT ANALYZER

ing technology and computing power. One common approach is to circumvent the bandwidth problem by employing specific knowledge about the task to reduce the data generated at each image. For instance, Nagle Research (Cedar Park, TX, USA) inspects the 3D surface of 250'000 pharmaceutical tablets per hour using a linear beam of structured laser light to only illuminate a profile of the tablet at each frame. The deviation from the reference line is used as a measure of the tablet's height. The disadvantage of such highly specialized solutions is the difficulty in adapting them to different applications.

A second strategy is to increase the effective bandwidth from parallelization of image acquisition and processing. For example, multiple temporally and spatially synchronized image acquisition systems have been used to increase the effective frame rate [Wilburn et al., 2005, Schuurman and Capson, 2004]. System bandwidth is increased only in proportion to the number of employed vision systems, however. The advantage of multiple camera systems is particularly problematic for small-scale applications, where space constraints are inherent.

Third, the bandwidth problem can be overcome by integrating image computations at the pixel level [Cembrano et al., 2004, Oku et al., 2005, Kagami et al., 2006, Miao et al., 2007]. Massively parallelized pixel level computations, such as linear convolutions [Cembrano et al., 2004], have been implemented in VLSI (Very Large Scale Integration) vision chips. In this way, relevant information is already computed on chip and only the higher order information needs to be transferred for further processing. An interesting variant of this concept has been implemented in a temporal contrast vision sensor that transmits the addresses (i.e. pixel locations) only of pixels experiencing supra-threshold variation in relative intensity [Zhang and Liu, 2006, Lichtsteiner et al., 2006]. By filtering relevant information content at the hardware level, only relevant image information is transferred with sub-ms latency to the image processing software, thus reducing the transfer requirements dramatically. VLSI techniques are bound to go beyond proof-of-concept experiments and will be be increasingly integrated in complex applications as their pixel array sizes increase and as they become more available as packaged products.

### 4.2.4 Design requirements

A high speed vision system was required to obtain a time resolved and spatially detailed concurrent measurement of both wing positions. We also desired a real-time read-out of the stroke parameters to control external hardware. For example, to implement a visual flight 'simulator', we intended to use the measured kinematic pa-

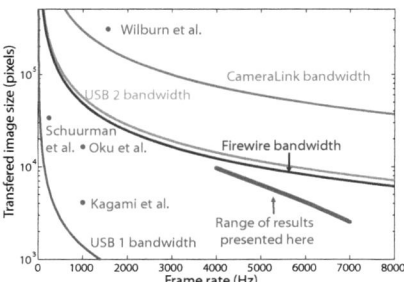

**Fig. 4.10:** Comparison between current high speed vision systems

rameters to provide the fly with realistic visual feedback in real time, as normally experienced in free flight. Similarly, we wanted to simulate realistic force feedback stimulation to specialized gyroscopic organs (halteres), which required phase-coupling the mechanical stimulus with the wing stroke [Nalbach, 1991]. The challenge, therefore, was to perform a real-time measurement of the complex movements of the fruit fly's two tiny wings (each only about 3 mm long), which beat back and forth more than 200 times each second.

To faithfully reconstruct the wing motion and identify relevant kinematic control parameters, the spatial resolution had to be sufficient to distinguish the subtle changes in wing kinematics, but coarse enough to allow the camera to run at sufficiently high framerates and for both wing paths to stay within the global FOV.

To determine the required spatial resolution we turned to earlier studies of flight control. The wing kinematics of tethered and free flying fruit flies were captured by Fry et al. at a frame rate of 5 kHz (about 25 samples per wing stroke) [Fry et al., 2003,Fry et al., 2005]. A semi-manual outline fitting procedure was applied to measure the time course of wing motion ( [Fry et al., 2003], their figure 1.B). In free flight, the changes of stroke amplitude occurring during fast turning maneuvers (body saccades) were measured in the range of 10°. We therefore opted for an angular resolution of 1°, which is sufficient to distinguish even subtle steering commands. Given a wing span of about 3 mm, the spatial resolution corresponded to about 26 µm in the world coordinate frame[2].

The global FOV needed to be large enough to cover the wing stroke path on each

---

[2]These calculations assume that wing position is measured halfway down the wing, and that the FOV encompasses the full wing.

## 4.2. DIGITAL WING BEAT ANALYZER

**Fig. 4.11:** Calibration procedure a) raw sequence images (1024x1024 pixels) b) statistical analysis c) area segmentation extraction of body and wings d) wing path extraction: the circular arcs represent the extracted wing paths

side, which required a FOV of at least 7x7 mm, or, given the spatial resolution above, 270x270 pixels.

Within the constraints of the system bandwidth, which in our case were limited by the camera, we were able to obtain sampling rates between 4 and 7 kHz, corresponding to 10 to 17 samples per stroke for each wing. The latter value represents an eight-fold increase to using a static ROI with the same hardware.

### 4.2.5 Implementation

We captured images of the tethered flying flies using a MVD-1024-Trackcam digital high speed video camera (Photonfocus AG, Pfäffikon, Switzerland), equipped with a 1024 by 1024 pixels CMOS image sensor. At full resolution, the camera achieves only 75 fps, however, the frame rate increases in proportion to a decrease in the chosen image size as predicted by (4.1) for a constant BW. The ROI and exposure settings could be updated on a frame by frame basis. The camera was connected to a Silicon Software (Mannheim, Germany) Micro Enable III Framegrabber via a Camera Link interface. We used a standard commercial PC with a Intel Pentium IV Dualcore 2.8 GHz processor and 1024MB of RAM. The operating system was Windows XP. To avoid system interrupts from causing delays (typically on the order of 1 ms), the program's thread was assigned real-time priority and ran separately from the operating system's threads on one of the dual processors, thus achieving real-time performance. The software was programmed using Visual C++ 6.0 and made use of OpenCV and Intel Performance Primitives (IPP).

The experiments were performed on an optical table. The camera was equipped with an Edmund Optics (New Jersey, USA) VZM 300i zoom lens with a primary magnification set to 0.75:1. The aperture was fully closed to 1.5 mm to maximize focal depth. The camera's exposure time was set to 50 µs to minimize wing blurring [Fry et al., 2003]. Due to the small aperture and short exposure time, extremely bright lighting conditions were required. We achieved this by backlighting with a randomized bundle fiber light guide attached to a 150 W halogen light source (Schott ACE, Mainz, Germany). Diffusive tracing paper was placed 15 mm in front of the light source to provide a homogenous light distribution. A 650 nm long pass filter (RG 715 Longpass, Edmund Optics) eliminated light in the range visible to the flies [Juusola and Hardie, 2001] to avoid behavioral artifacts.

Flies were immobilized at 5 °C on a custom built Peltier cooled stage and glued to a tungsten probe with UV cured glue (Loctite, Duro Clear Glass Adhesive) using standard techniques [Tammero and Dickinson, 2002]. The probe with the attached fly was positioned along six degrees of freedom with a micromanipulator (Sutter MP285), such that the stroke plane roughly coincided with the camera image plane and the wing stroke path was not occluded by the tether.

### 4.2.6 Calibration

ROI based wing tracking and an optimized analysis of the sampled images requires detailed knowledge of the wings' expected paths at the time of tracking. Because the tethering and lighting conditions vary considerably between consecutive preparations with different flies, detailed image information needs to be acquired between consecutive measurements. For this, we employ an automatic calibration procedure that a) segments the FOV into functionally relevant areas containing the fly body, the tether and the wing stroke envelope; b) determines the optimal threshold to distinguish the wings from the background; c) determines the pixel paths along which the wing will be followed during the tracking.

**Image segmentation**

The algorithms for image segmentation are based on an analysis of the varying image characteristics between functionally distinct regions (body, tether and stroke envelope). First, a set of 100 full-frame (1024*1024 pixels) images are acquired at 75 Hz,

## 4.2. DIGITAL WING BEAT ANALYZER

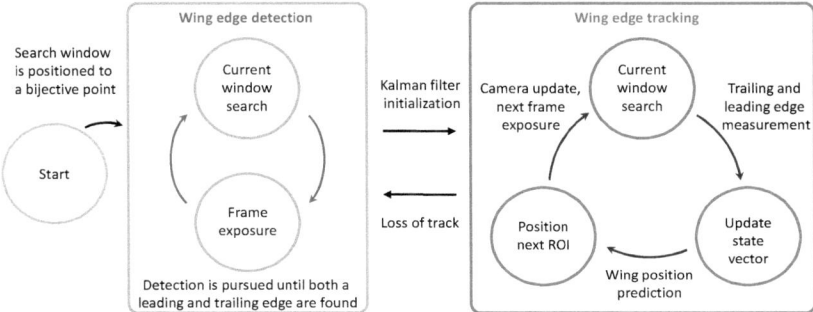

**Fig. 4.12:** Tracking state machine

whereas the wing phase in each image is arbitrary (see Fig.4.11a). Next, a statistical analysis is performed of the pixel values throughout the sequence. This results in two images (see Fig.4.11b). The first defines a mask for the stroke envelope based on a binarized image of the variance. The second defines a background image based on the median pixel values.

The dark halo caused by the background lighting is then removed using a vignetting filter and the tether is extracted using a Hough transform. Finally, the body blob is extracted from size and position criteria and its orientation and heading used to separate the left and right sides of the fly (Fig.4.11c Body).

### Wing segmentation characteristics

The appropriate choice of the threshold used to track the wing edges during tracking is critical. To accommodate for the variation between individual fly preparations, the wing edge detection threshold also needs to be calculated during the calibration phase.

The threshold is determined in two steps. First, the background image is subtracted to increase the signal-to-noise ratio by rendering the system robust to spatial variations in back-lighting. Second, the Max-Lloyd quantization algorithm is applied to the pixels belonging to the wing envelope (white pixels in the variance image Fig.4.11b). This results in an optimal threshold in the least square's sense (see Fig.4.11c Wing).

**Wing path extraction**

The wing tracking paths are the key information extracted during calibration. For each wing, a circular arc with an optimal radius and centered on the wing hinge is chosen. These arcs correspond to a list of pixel locations used to locate the wing edges and position the ROI in subsequent frames.

To obtain the wing hinge position of each wing, the wing envelope is binarized using the wing edge detection threshold (see above), which provides a representation of the isolated wing. A morphological opening is performed to eliminate possible holes within the wing blob caused by the semi-transparent sections of the wing. Next, a Canny edge detector is applied to isolate the wing edges (see Fig.4.11c Wing edge). A Hough transform is applied to extract the strongest line. In most of the cases, this line corresponds to the leading edge of the wing. The intersection between each strongest line of the sequence is calculated. The median intersection is taken as an estimate of the wing hinge position (marked with full circles in Fig.4.11d).

Combining the knowledge of the wing hinge positions and the stroke envelopes, the optimal circular path segments for subsequent wing tracking are chosen by applying an angular range criterium (see the circular arcs in Fig.4.11d). The sequence of pixel locations of the path segments are stored together with the corresponding value describing the angle subtended with respect to the body, providing a lookup table for wing position in angular coordinates of the body.

### 4.2.7 Real-time data acquisition

To track the wings robustly, the ROI needs to be placed in appropriate positions along the wing path and the wing position in the ROI determined using efficient image analysis. To initiate and perform wing tracking, as well as to recover from loss of tracking (e.g. because of flight interruption), we implemented the tracking state machine shown in Fig.4.12. The process is described in detail below.

**Wing edge detection**

The tracking process begins with a detection mode, in which the ROI is placed at one of the two limits of the wing path (maximum or minimum point). This point has two advantages; first, because it is bijective, it corresponds to a unique phase of the wing beat kinematics, creating an unambiguous starting point to the tracking system. Second, the low wing velocity at these points allows the parametric model to adjust itself without losing track. Images are continuously acquired and analyzed

## 4.2. DIGITAL WING BEAT ANALYZER

using the static ROI until both the leading and trailing edge of the wing are detected. The parametric model is then initialized and the state machine switches to the wing edge tracking mode.

**Wing edge tracking**

In the wing edge tracking mode, the ROI is dynamically positioned according to the prediction of a parametric model. The ROI is placed at the position of the expected wing traversal, which can be several wing chords distant from the previous measurement. The current ROI is grabbed and the image processed (see below) to obtain a measure of the wing position, which is then used to update the parametric model of the wing kinematics. The ROI is then placed according to the prediction of the updated model and the process is repeated. If the process fails to detect the wing edges, it continues to estimate the ROI position according to its last measurement. After about 500 µs, the system reverses to the detection mode and sets the ROI back to its initial position. The grace period gives the system a chance to recover from faulty predictions, which typically happen during the first wing beat after the Kalman initialization. The longer interruptions were usually the result of an interruption of flight.

**Image analysis: wing extraction**

The time budget allotted to the processing of the individual images does not allow CPU intensive image analysis, such as blob analysis and pattern matching, to be applied. Using instead only the circular paths extracted during calibration (see 4.11), it is possible to reduce the number of pixels to be processed from 2500 to about 60 and perform the necessary image analysis within a time frame of about 50 µs (see ?? and Fig.4.16 for details).

After transferring the image from the camera to the frame grabber, the following procedure is used to detect the wing edges: The pixel values of the wing path are copied from the frame grabber's onboard memory to an array in the computer's RAM (Fig.4.13.1). The background image captured during calibration is then subtracted from the sequence (Fig.4.13.2) and the resulting sequence binarized using the wing detection threshold (Fig.4.13.3). A morphological opening is applied to fill in gaps caused by the semi-transparent sections of the wing (Fig.4.13.4). Finally, an edge detector extracts the leading and trailing edge of the wing (Fig.4.13.5).

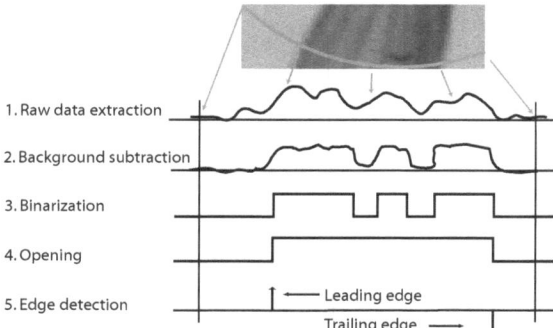

**Fig. 4.13:** Edge extraction algorithm applied along the wing path during tracking. The binarization threshold is based on the Max-Lloyd quantization performed during calibration. The morphological opening width is manually chosen to remove objects that are the size of the legs, but not objects that are the size of the wings.

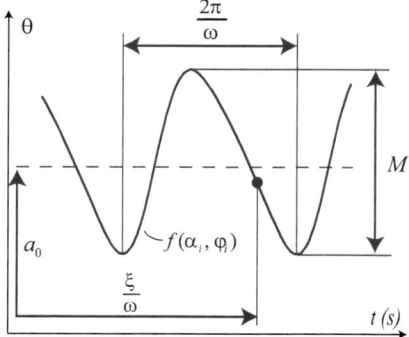

**Fig. 4.14:** Significance of the state vector: the shape of the model is defined by the relative amplitudes $\alpha_i$ and relative phases $\omega_i$ of each Fourier component ($N = 3$). This shape is scaled and offset, both in time ($\omega$ and $\xi$) and space ($M$ and $a_0$) by the EKF to fit the measurements.

## 4.2. DIGITAL WING BEAT ANALYZER

**Fig. 4.15:** Sample from a tracking sequence, the squares and circles represent the extracted positive and negative edges, respectively. The rectangles show the borders of the ROI inside of which pixels were exposed and transferred during a given frame. The median image extracted during calibration is shown in the background for ease of interpretation.

**Process model: implementation of an Extended Kalman filter**

The parametric model used to predict wing motion was implemented using a Kalman filter, by virtue of its computational efficiency (due to its recursiveness) and optimality with respect to most statistical criteria [Maybeck, 1979].

The Kalman filter is initialized prior to entering the wing tracking mode. Its initial state estimate is based on parameters extracted during calibration and on previous knowledge of wing kinematics, e.g. typical wing beat frequency. During tracking, four Kalman filters are used simultaneously to track the positive and negative edges of both left and right wings. The measured wing edge positions are used to update the state estimate and thus increase the accuracy of the prediction for the subsequent measurements. To position the ROI in the subsequent frame, the current state estimates are used to predict the future positions of the positive and negative edges of the wing. The ROI is then centered around the mean value. As a further kinematic parameter, the difference of the wing edge coordinates provide a measure of the wings' angle of attack.

Because the Kalman filter state estimate represents an online parametrization of the stroke kinematics, it can be used to control external processes in real time. An example for its use as a behavioral metric for real time feedback control is given in section 4.2.8.

The periodic motion of each wing edge is modeled using a Fourier series:

$$\theta = a_0 + \sum_{i=1}^{N} a_i \cos(i\omega t) + b_i \sin(i\omega t) \tag{4.3}$$

where $\theta$ represents the 1D angular position of the edge with respect to the hinge and N is the number of Fourier terms.

By constraining all parameters except amplitude, offset, frequency and phase, we maintained the original shape of the function, while scaling it in time and space. (4.3) is rewritten to incorporate these state variables (also see Fig.4.14):

$$\begin{cases} \theta = a_0 + \sum_{i=1}^{N} \alpha_i M \sin(i\xi + \varphi_i) \\ \xi = \omega t \end{cases} \tag{4.4}$$

where $\alpha_i$ is the relative amplitude of the Fourier coefficient $i$, $M$ is the amplitude of the function, $\xi$ contains the phase information of the function with respect to the frequency $2\pi\omega$ (phase is dependent on frequency), and $\varphi_i$ is the relative phase of the Fourier coefficient $i$. The $\alpha_i$ and $\varphi_i$ coefficients are static values describing the shape of the model. The state vector is $x_k = \begin{bmatrix} a_0 & M & \xi & \omega \end{bmatrix}^T$, and the derivation of the discrete state dynamics from (4.4) yields:

$$x_k = \underbrace{\begin{bmatrix} 1 & 0 & 0 & 0 \\ 0 & 1 & 0 & 0 \\ 0 & 0 & 1 & T_s \\ 0 & 0 & 0 & 1 \end{bmatrix}}_{A} x_{k-1} + W \tag{4.5}$$

$$W \sim N(0, Q)$$

where $T_s$ is the discrete time step, $A$ relates the state at the previous time step to the state at the current step, $W$ is the process noise and $Q$ is the process noise covariance. This noise is introduced to take into account the changes in kinematics performed by the fly (which to measure is our goal) and the imprecision inherent to any model.

The measurement equation is nonlinear. Therefore, an Extended Kalman Filter (EKF) is used, which linearizes the measurement equation at each time step:

$$\begin{aligned} \theta &= H\, x_k + v_k \\ H &= \begin{pmatrix} \frac{\partial \theta}{\partial a_0} & \frac{\partial \theta}{\partial M} & \frac{\partial \theta}{\partial \xi} & \frac{\partial \theta}{\partial \omega} \end{pmatrix} \\ v_k &\sim N(0, R) \end{aligned} \tag{4.6}$$

where $v_k$ is the measurement noise and $R$ is the measurement noise covariance. In the experiments, the static parameters $\alpha_i$ and $\varphi_i$ with a total of three Fourier terms ($N = 3$) were chosen to fit the kinematic measurements of the wing beat presented

## 4.2. DIGITAL WING BEAT ANALYZER

in [Fry et al., 2005]. The two parameters of the Kalman filter, $R$ and $Q$, were determined empirically from a first series of measurements. The components of $Q$ affect how much the state variables will vary given a new measurement. These weights must be chosen to achieve a correct sensitivity. $R$ represents the confidence that a measurement is correct (not a false-positive). Together, $R$ and $Q$ determine how the model parameters will get affected by the new measurement: the state vector estimate is updated for each measurement by solving the Ricatti equations.

**Process time line**

The different processes must work together efficiently in a synchronized way to achieve the required frame rates given the hardware constraints. During wing tracking, three pieces of hardware are running in parallel: the camera, the frame-grabber and the computer's CPU. These three components must be able to determine where the ROI must be placed before the next frame is transferred. The camera (Fig.4.16, top row) initially exposes the image during 50 µs. The camera then transfers the image to the framegrabber (Fig.4.16, middle row). The computer receives a signal once the transfer is complete (Fig.4.16, bottom row) and starts the image processing. The wing position is extracted as described in Section 4.2.7. The Kalman filter uses the measured position to update its state vector and the following ROI position is calculated (see previous section). This ROI is sent to the framegrabber which updates the camera once the current frame has been transferred.

Although the process for a single frame is serial, up to three of these processes are running simultaneously on the different hardware components to maximize the speed. For instance, the next frame is exposed and transferred while the current frame is being processed. This does not affect the tracking performance because the left and right sides are exposed sequentially, giving an extra frame period to update the ROI on each side.

**Data logging and hardware control**

During tracking, the Kalman filter state vector, the position of the ROI and the position of the leading and trailing edge of the wings are stored in the computer's heap for each captured frame. The camera data (ROI pixel values) are left on the frame grabber where they are overwritten during the subsequent buffer cycle. While this procedure does not retain the majority of the image data due to bandwidth requirements, the last 1GB of data remain available on the framegrabber and were retrieved and saved to disk for post processing (Fig.4.15, see also Discussion).

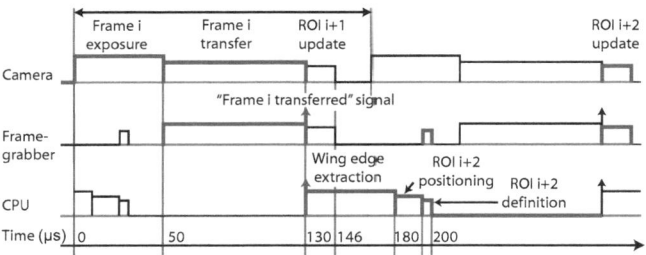

**Fig. 4.16:** Process time line: Procedures related to frame i are in bold.

### 4.2.8 Results

The resolution of the wing position was about 1° in our experiments. Various combinations of ROI sizes and frame rates were used to explore optimal parametrization for robust and detailed tracking. We obtained robust tracking performance in a range of FOVs and sampling rates between 4000 Hz and 7000 Hz, which was sufficient for our application.

In our system, the current limitation was the bandwidth of image acquisition and not image processing. At 7000 Hz, the size of the ROI is barely larger than the full width of the wing. Therefore, a further increase in temporal resolution could only be achieved at the cost of spatial resolution by reducing the magnification of the lens.

As shown with a representative example in Fig.4.17, we were able to measure the positions of the two wing edges reliably. Our data closely resemble earlier published data on stroke angle ( [Fry et al., 2005], their Fig.4.A).

We reconstructed the sampling procedure of ten consecutive frames based on data from the image buffer saved at the end of a measurement (See Fig.4.15). The sequence shows a half a stroke cycle starting with an early down stroke (image 1) to ventral stroke reversal (image 9).

Measurements up to 120 s, or 840'000 samples, were routinely performed. A representative example of such a measurement is shown in Fig.4.18. The fly was measured as it was optically stimulated with a vertically-oscillating pattern. The sinusoidal response of the fly is clearly visible.

Table 4.3 shows the tracking statistics collected with $n = 10$ flies. The robustness of tracking is high with over 80% of frames in which a full wing was extracted and

## 4.2. DIGITAL WING BEAT ANALYZER

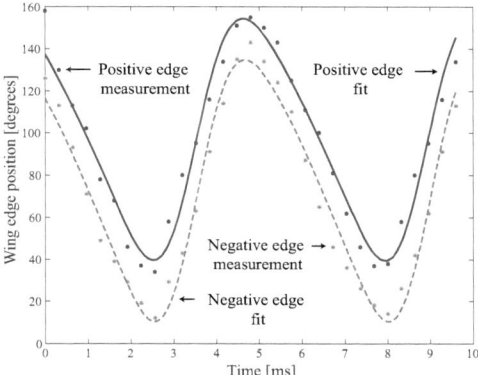

**Fig. 4.17:** Tracking results for one wing with an effective sampling rate of 6250/2 Hz. Positive and negative wing edge positions are marked with circles and stars, respectively. The positive and negative edge Kalman fits used to estimate wing position is represented with a solid and dashed line, respectively.

over 99% of time spent in "wing edge tracking" mode[3]. Occasionally, tracking fails during mid-stroke, where the wings move the fastest, but it typically recovers within the same stroke as the wing slowed down toward stroke reversal. The mistracking is caused by a mismatch between the fly's kinematics and the EKF's *a priori* model. To address this issue, we developed an offline algorithm that can adjust the EKF parameters $\alpha_i$ and $\varphi_i$ given an initial measurement set. Incomplete wing extraction occurs when both wings temporarily overlap during dorsal flip, and only one of the wing edges can be extracted. Furthermore, flies sometimes extend their legs into the wing path, which led to false wing position readings. These effects are all visible in the visual data we collect at the end of the measurement (see Fig.4.15) and can be avoided by a better calibration and an appropriate choice of the opening width in Fig.4.13.

---

[3]The initial tests required to get a proper calibration were removed from the data set. With an improper calibration, the tracking performance is poor.

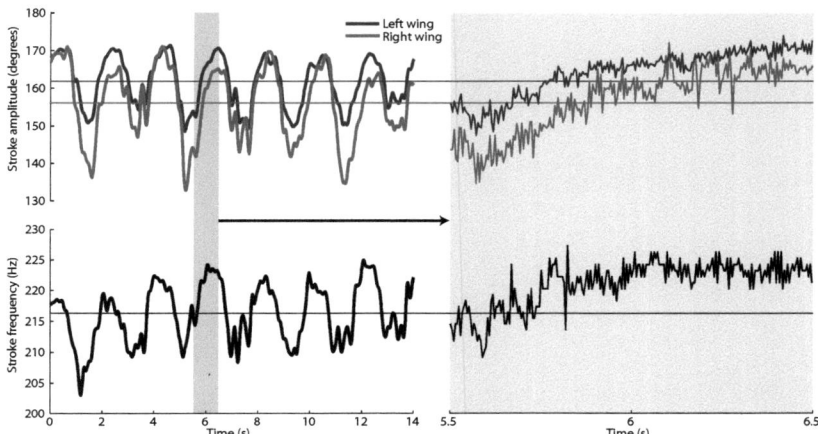

**Fig. 4.18:** Typical stroke amplitude and frequency measurements for a fly stimulated with a pattern oscillating vertically at 1 Hz. The fly responds by sinusoidally modulating its wing beat frequency and amplitudes. The measurements were first interpolated to precisely extract both stroke reversals of each wing beat. The amplitude and frequency were then extracted from these extrema. Data on the left hand side were low pass filtered with a Butterworth filter to improve visibility. The right hand side shows a one second extract of the unfiltered data. The horizontal lines represent the mean values throughout the measurement.

**Table 4.3:** Tracking statistics based on $N > 28 \cdot 10^6$ frames measured over $n = 10$ flies.

| Property | Value | Inter-fly S.D. | Unit |
|---|---|---|---|
| # flies | $n = 10$ | N/A | N/A |
| Total # of frames analyzed | 28'557'764 | $2 \cdot 10^6$ | frames |
| Total tracking time | 9142 | 644 | s |
| Successfully extracted both wing edges | 80.0 | 9.31 | % |
| Extracted one wing edge | 15.9 | 6.2 | % |
| No wing edge found | 4.1 | 4.8 | % |
| Time spent in "Wing edge tracking" mode | >99.9 | <0.1 | % |
| Mean prediction error of Kalman Filter | 6.07 | 1.77 | ° |

## Proof-of-concept experiments

As a proof-of-concept for the general applicability of our process model for real-time hardware control, we used the stroke phase estimate provided by the process model to trigger a stroboscopic flash at defined times during the stroke cycle on a stroke by stroke basis. By linearly varying the phase position between successive strokes, we generated a 'slow motion' impression of wing motion that could easily be verified by observation. An example of a stroboscopic video can be found in the supplementary material.

To control the strobe lighting, we estimated the wing phase after each Kalman update and calculated the remaining time to the subsequent strobe flash. If this time was calculated to be less than the camera's frame time, a digital message was sent out that contained information about the precise timing of the flash[4]. The timing information was decoded on dedicated hardware and used to trigger a flash of 50 µs duration. This application provided a live view of apparent slow motion that revealed induced kinematic changes from stimulation, but also problems occurring during tracking.

### 4.2.9 Discussion

**Comparison with existing technologies**

In previous research, several solutions have been proposed to overcome bandwidth limitations to perform real-time high speed tracking. Analog or hybrid analog/digital vision chips provide particularly intriguing solutions as they allow processing power to be integrated at the pixel level. Not only does this allow a massive parallelization of image pre-processing, but it also offers the opportunity of transferring image information selectively to increase the information transfer rate. While these technologies promise highly effective and affordable tracking solutions, implementations using standard digital and software development technology nevertheless offer several advantages.

First, standardized digital solutions have a high degree of flexibility to adapt to specific measurement or control requirements. In our solution, the parameters can readily be tuned to different experimental situations. Such flexibility is difficult to achieve in vision chips, in which a specific task is typically built into the hardware.

Furthermore, a graphical representation of the tracking process is maintained in

---

[4]The number of µs until the actual flash was encoded by seven bits and sent through the computer's parallel port with a total delay of about 20 µs.

digital systems, in our case in form of the final 1GB of acquired image data. The availability of raw visual data allows additional off line analysis to be performed e.g. as a control for meaningful position extraction. The analysis of the tracking process in analog systems, on the other hand, can be challenging and hard to interpret.

**Concept generalization**

Our contribution can be summarized as the combination of three key concepts that together form a novel solution to high speed tracking: 1. Dynamic region-of-interest sampling 2. Real-time image processing and 3. Online process model.

1. **Dynamic region-of-interest** Our implementation in digital hardware makes use of the dynamic region-of-interest capabilities offered by recent CMOS based digital video cameras. The benefit from transferring only a small part of a FOV is of general relevance to tracking applications in which the bandwidth gain is worth the extra processing time involved with the prediction and update of the ROI. This is the case if the object of interest is small compared to the global FOV.

    As sensors increase in pixel array size and frame rates, the bandwidth problem will certainly become more pronounced, making dynamic region-of-interest approaches more appealing. The concept of dynamic region-of-interest need not be limited to standard CMOS cameras. A scanning electron microscope (SEM), for example, can easily be configured to scan a specific area, making it ideal for a dynamic region-of-interest approach in time-critical cases. This is especially relevant because objects at small dimensions tend to move fast with respect to their size.

2. **Real-time image processing** The real-time image processing is inherent to our solution because the region of interest must be updated as a function of the current wing's position. A given image had to be analyzed before the following one was completely transferred. By analyzing each image online, only the useful information has to be stored. As a consequence, the system consumes very little memory and can therefore run almost indefinitely. This is useful for applications that must test large parameters spaces, or that are waiting for a single event to occur.

    To analyze each image within a few microseconds, we benefited from some simplifying assumptions inherent to our application: the fly was tethered, allowing us to extract the paths of the wings and only analyze relevant pixels and therefore drastically reducing the computation costs. Similar simplifications can be found in many tracking applications: higher sampling rates decrease the distance an object has moved between two frames, such that the search of an object can first concentrate on the pixels most likely to contain it, and then extend to pixels with lower likelihoods. Limiting the analysis to a small amount of pixels allows even complex signal processing computations to be performed fast.

3. **Online process model** To be able to position the region-of-interest, the position of the object must be predictable from past measurements. The prediction model is therefore a key element of our tracking approach. Here, a tradeoff must be found on the level of complexity of the model. A more complex model will predict the position of the object more accurately, therefore allowing a smaller region of interest to be chosen and increasing the frame rate of the camera. A more complex model also involves more computations , however, and will add time to the processing. The level of complexity must therefore be chosen in an iterative way to find an optimal tradeoff between transfer and processing constraints.

In our application, we opted for an Extended Kalman Filter mainly due to its recursive implementation and its ability to precisely extract a state vector at each measurement instance, therefore providing real-time analysis of the measurement time-course. Other similar techniques, such as alpha-beta-gamma filters, do not provide a real-time analysis of the data.

## 4.2.10 Concept generalization

This real-time functionality is crucial to applications that require the control of an external piece of hardware based on its current process state. This is of general relevance to behavioral research paradigms in biology, because it provides controllable sensory conditions that nevertheless depend on the animal's (intended) behavior in a naturalistic way. As robotics become more autonomous and integrated into processes, the need for rapid and robust process estimations based on non-contact sensors will also increase. For instance, production lines are likely to increase in speed and complexity. Likewise, visual servoing will be applied to faster manipulators. In these cases, the complexity and structure of the process model is completely dictated by the required feedback.

In our application, two of the Kalman filter's state variables were directly employed for the control of external hardware. Because the difference in stroke amplitude between the right and the left wings are associated with intended turning maneuvers of the tethered fly [Dickinson et al., 1993], we could use them to control the fly's visual panorama in a 'flight simulator' [Heisenberg and Wolf, 1988].

We also used the Kalman filter's phase variable to stroboscopically illuminate the wings or excite the mechanosensors at precisely defined wing phases.

The EKF presented in this paper can be applied to any periodic motion. The shape of the fitted curve is entirely defined a-priori and is then scaled and offset both in time and space during the tracking, providing an estimate of the current amplitude, phase, frequency and offset of the periodic signal. For non-periodic motions, the EKF can easily be adapted to, for example, a constant acceleration model that predicts the future location of the object based on its present position, velocity and acceleration.

The combination of dynamic region-of-interest, real-time image processing and online process model promises broad applications in experimental research and process control, whenever the status of a system must be rapidly evaluated as part of a control loop.

## 4.3 Visual flight simulator

Up until now, we have described tools to measure the output of the fruit fly: the wing kinematics and the flight forces. These tools can be used on their own, but to truly release their power, they must be used in combination with an input to the fly. The measurements then represent the response of the fly to a known stimulation.

In tethered flight, most of the fly's sensory systems do not receive natural feedback because the body is fixated. The sensory input must then be simulated. Here, we present such a simulator that provides input to the fly's most powerful sensory modality: vision.

### 4.3.1 Other visual flight simulators

Presenting visual patterns to tethered insects has been done for quite some time. Initial experiments used patterned cylinders that were mechanically rotated. This experiments led to important discoveries of basic sensorimotor principles in insects, such as the principle of reafference [Von Holst and Mittlestaedt, 1950], the Hassenstein-Reichardt correlation model [Reichardt, 1961] and the optomotor yaw turning response [Götz, 1964]. Due to the limited flexibility of the system, researchers then

## 4.3. VISUAL FLIGHT SIMULATOR

developed projector based systems. Goetz *et al* used two independently controlled projectors for each eye [Gotz, 1968]. Hausen *et al* used pattern projectors mounted on a gimbal [Hausen, 1982]. Krapp *et al* developed a system where a small dot could be positioned at different viewing angles [Krapp and Hengstenberg, 1997]. Afterwards, several groups developed custom built LED solutions [Lehmann and Dickinson, 1997, Strauss et al., 1997, Lindemann et al., 2003], that provided more flexibility in the pattern choice as well as higher sampling rates. An improved, modular, LED panel system was developed by Reiser *et al* [Reiser and Dickinson, 2008], and the design was released as an open source project. At the same time, interest in projection-based systems has gone back up in recent times because of the emergence of faster DLP chips and the possibility to project on spherical domes [Taylor et al., 2008].

### 4.3.2 Design requirements

The basic requirement is to have a spatial resolution fine enough and a temporal resolution fast enough so that the fly doesn't see the increments. These would produce artifacts. The fly's interommatidial angle is about 4° [Götz, 1964] and the flicker fusion rate was found to be inferior to 150 Hz [Miall, 1978]. The field of view should cover as much of the flies natural field of view, which is almost panoramic. The system has to produce light at wavelengths visibles to fruit flies. Heisenberg *et al* showed that the relative response of *Drosophila* was highest for wavelengths between 350 and 500 nm [Heisenberg, 1977]. The system should also be able to cover as much of the fly's luminosity range as possible. Such a measure is difficult to perform in absolute terms and it is usually considered sufficient if the behavioral response to the visual input saturates as of a certain intensity.

### 4.3.3 Implementation

Given the large number of previous flight simulator implementations, it would have been inefficient to design our own from zero. We chose to base our simulator on the LED system designed by Reiser [Reiser and Dickinson, 2008]. This design was the only open-source one and fulfilled the specifications described above. Furthermore, we had direct contact with M.B. Reiser, making the transfer of know-how easier.

After a few months, however, we noticed certain limitations of the system in terms of the data rate, which, in certain cases, was as low as 15 Hz. At the heart of this problem was the Two Wire Interface (TWI) that was used to communicate from a single master to a total of 96 slaves. We decided to remove this bottleneck

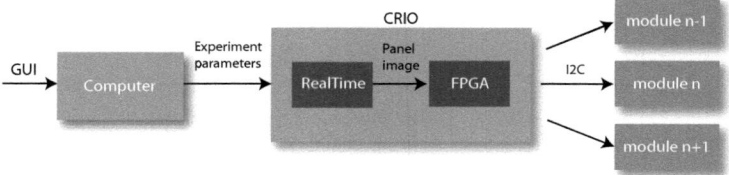

**Fig. 4.19:** Concept of panel system. The user interacts with the computer through a LabVIEW GUI. The computer sends over to the CRIO the spatiotemporal parameters of the experiment. During the experiment, the Real-time controller of the CRIO generates images that are transformed into I2C signals by the FPGA. The I2C signals are addressed to individual modules, which each show a 8x8 pixel pattern.

by implementing a novel controller that used twelve TWIs to communicate with the slaves. Apart from this and a few other small improvements, the system we used was a copy of the design by Reiser.

Our system is composed of three parts: i) a computer, where the user can specify the type of experiment and the visual parameters. The computer also logs the information of different events, such as the synchronization with external hardware or the image publication characteristics. ii) a real-time controller, which generates each image, transforms it into individual instructions for each panel and sends these instructions to the panels. iii) the individual panel modules, each consisting in a 8x8 LED matrix and a 8bit microcontroller. The different panels are assembled together to form a cylindrical shape.

**GUI computer interface**

The computer contained a custom designed GUI where the user could easily select the general type of experiment (Open-loop dynamic stimulations, static images, closed-loop stimulations), and the different visual parameters, such as the temporal and spatial frequencies of a sine grating.

Once the experimental parameters have been chosen, and the experiment has be launched, the computer sends the information over to the real-time controller. At the end of the experiment, this one will return important information from the experiment. The program will log the information into text files for future data analysis.

## 4.3. VISUAL FLIGHT SIMULATOR

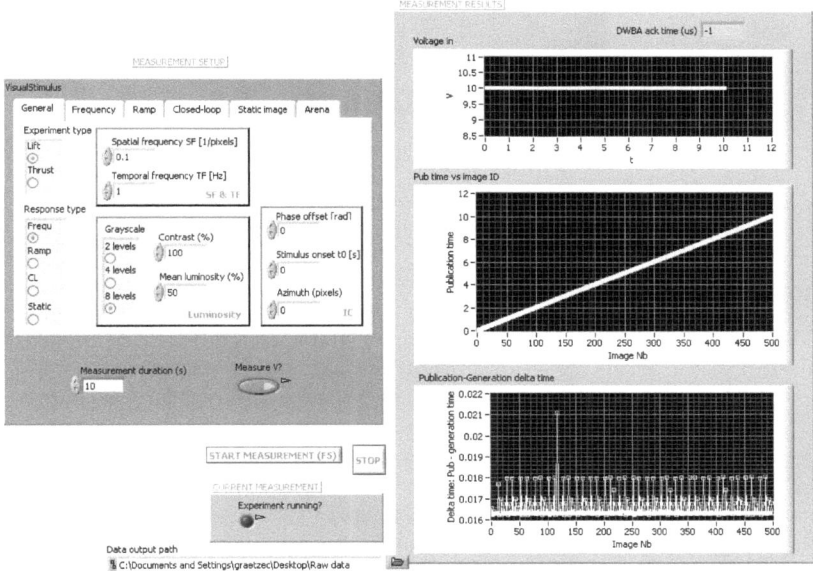

**Fig. 4.20:** LabVIEW GUI of the panel system. The user can choose the spatiotemporal parameters of an experiment, launch it and observe the measurements.

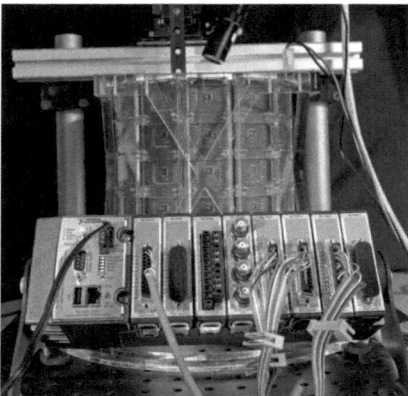

**Fig. 4.21:** CompactRIO system used to generate patterns and send them off to the modules.

**Real-time controller**

The real-time controller runs in the background and the user never has to directly interact with it. It contains a CPU running VxWorks (www.windriver.com/vxworks/) and a FPGA backplane that allows to parallelize most of the low level interfacing.

The real-time controller first waits for instructions from the computer. Once it receives the experimental parameters, it starts by generating the image to be shown on the panel. The image is then segmented into the 8x8 pixels of each panel module and the I2C command for that panel is generated. The I2C commands are sent out on 12 parallel lines - one for each vertical row of panels. This significantly increases the speed of the system compared to the original design by Reiser. The real-time controller receives the image publication time stamps, and will pass this information back to the computer. The real-time controller also logs the force data from the single axis force sensor and synchronizes itself with the Digital Wing Beat analyzer, so that the timestamps of both systems are directly comparable.

**LED modules**

The LED modules consist in a microcontroller, that is programmed to receive the I2C commands and transform them into control signals to the LED, via a set of Darlington transistors. To achieve 8 different greenscale values, the LEDs are refreshed at

## 4.3. VISUAL FLIGHT SIMULATOR

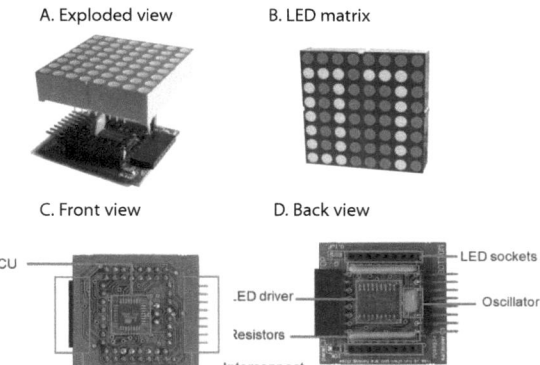

Fig. 4.22: Overview of an individual module. A: The modules connect to LED matrices. B: Front view of a LED matrix. C: Front view of module PCB. D: back view of module PCB.

approximately 2600 Hz in a cycle of 8 ON/OFF events, e.g. a pixel set to a greenscale of 4 will be on during 4 of the 8 cycles. The worst-case refresh rate, 372 Hz, is well above the flicker fusion rate recorded for fruit flies [Miall, 1978].

#### Completed arenas

We built two arenas, the flat and the circular arena. The flat arena used 96 individual LED modules and was designed to emulate conditions in the free-flight wind tunnel. The circular arena used 60 individual LED modules and was designed to cover the entire panoramic range of the fly.

#### Performances

By parallelizing the I2C lines, we were able to increase the performance of our system. It achieves 50 Hz with a non-symmetrical, grayscale image, 150 Hz with non-symmetrical black/white images and over 250 Hz for images with vertical symmetry.

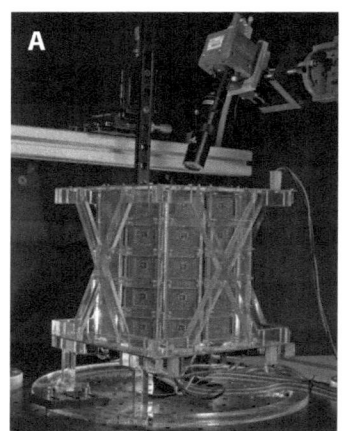 

**Fig. 4.23:** A: Circular arena  B: planar arena

# Chapter 5

# Reverse-engineering micro-aerial flight in biology

I applied the technologies described in the previous chapter to a set of different biological problems. These were based on some of the questions relative to flight control which were discussed in the introduction (see Chapter2) and which were compatible with our technologies.

**Table 5.1:** Overview of the different projects and their dependency on the developed technologies

|  | DWBA | MEMS 1 DOF | MEMS 3 DOF | Planar LEDs | Circular LEDs | Piezo stim. |
|---|---|---|---|---|---|---|
| Biomechanics of tethered flight, §5.1 | X | X |  |  | (X) |  |
| Lift control, §5.2 | (X) | X |  |  | X |  |
| Thrust control, §5.3 | (X) |  | X | X |  |  |
| Role of mechanosensors | X |  |  |  | X | X |
| Exploration of bio/robot coupling, §C | X |  |  |  | X |  |

Table 5.1 describes the main classes of experiments and their technological requirements.

The biomechanics of tethered flight studied the relationship between the measured kinematics and forces. This relationship was compared to an aerodynamic model that was previously extracted using dynamically-scaled robotic models of the wing (see Section 5.1, page 56).

The study of lift control analyzed the equilibrium reflexes of flies for lift (altitude) control. In this study, the fly's response was characterized in the same way as an artificial controller (see Section 5.2, page 59).

The study of thrust (forward velocity) control aimed at comparing the tethered data with free-flight experiments performed in a wind tunnel (see Section 5.3), page 68).

The role of mechanosensors was studied by Jan Bartussek in a collaboration that used the developed setup.

Finally, the bio/robot coupling experiments explored the transfer of a biological control strategy into a robotic implementation. To this end, the fly's output were coupled with the steering commands of a two wheeled robot, and the robot's sensor data was fed back to the fly visually (see Section C).

Taken together, these applications show the breadth of biological applications made possible by the developed technologies. This list corresponds to the initial projects that could be undertaken within the time frame of this thesis. There are many more experimental paradigms that could be measured. I will now go through each topic in more detail.

## 5.1 Biomechanics of tethered flight

In the past decade, the increasing interest of roboticists to build artificial micro-flappers as well as the availability of novel tools has brought a lot of interest to the lift generation mechanisms of micro-aerial insects. These studies mainly used the dynamically scaled robotic wing approach (see 2.1.3). These studies identified a set of models of how the fruit fly generates lift [Dickinson et al., 1999, Sane and Dickinson, 2001, Sane and Dickinson, 2002, Birch and Dickinson, 2001], however, a validation of these principles using direct force measurements was never performed. we therefore set off to validate these, together with Vinzenz Schönenfelder, who did his Master thesis under our supervision.

To validate the model, we concurrently measured flight forces with the MEMS force sensor, and wing kinematics with the DWBA. The wing kinematics extracted from the DWBA were plugged into the "Revised quasi-steady model" which is the most detailed model available at the moment (see [Sane and Dickinson, 2002] and Section 2.1.3). We then compared the model's predictions with the measured forces.

**Revised quasi-steady model**

Combining aerodynamic theory and empirical results from a robotic wing model, Sane and Dickinson proposed a revised quasi-steady model [Sane, 2003, Sane and Dickinson, 2001]. Besides steady-state forces, the model additionally included dynamic effects resulting from wing acceleration and rotation. Skin friction is neglected.

The revised quasi-steady model represents aerodynamic forces as the sum of three independent components:

1. **Translational force** corresponds to the standard quasi-steady model and is equivalent to the steady-state forces on airfoils.

2. **Rotational force** derives from the Magnus effect. [Dickinson et al., 1999]. The force emerges when a translating wing is rotating about its span-wise axis.

3. **Added (or virtual) mass force** arises from the inertia of the surrounding air [Ellington, 1984b]. It only acts when the wing accelerates (dragging the surrounding air with it) or rotates (which also results in an acceleration of the surrounding air).

A major shortcoming of the revised quasi steady model is that it is only implicitly dependent on time, i.e. it only allows for the time-dependence of the wing kinematics, but not of the actual fluid flow. Thus, it neglects the aerodynamic forces resulting from wake capture, which is considered a significant component of force production in *Drosophila* [Dickinson et al., 1999, Sane and Dickinson, 2002, Wang et al., 2004].

The details of the model implementation are described in ChapterB.

## 5.1.1 Results

The modeled and measured forces were quite similar (see Fig.5.1). The magnitude of the forces are the same and the mid-stroke components match quite well. At stroke reversals, the modeled and measured forces differ. Here is a list of possible causes, with the most probable first:

- The revised quasi-steady model provides only a simplified estimation of aerodynamic forces. We had to omit the added mass component, we also neglected the inertial effects of the air surrounding the wing (see Section B.3). There was, however, no alternative, since this has been the most precise model published from the dynamically scaled robotic wing experiments.

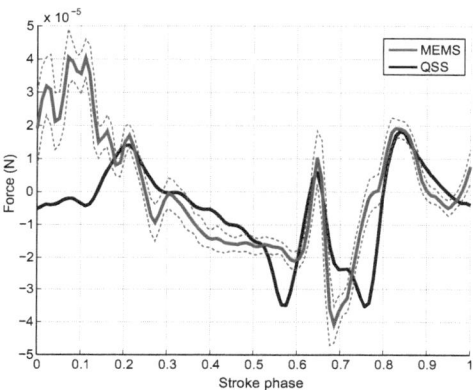

**Fig. 5.1:** Comparison of modeled vs measured lift forces. The model details are described in ChapterB.

- The wing beat analyzer can only provide the position of the wing in the camera plane. We extrapolated the full three dimensional movement of the wing by superposing the measured data with previously published wing kinematics [Fry et al., 2005]. This extrapolation obviously affects the precision of the model.

- All constants characterizing the insect's morphology were not directly determined, but assumed to range in standard values found in the literature (see Section B.1). Wing mass, surface area and other morphological parameters all influence the instantaneous flight force production. Estimating these values by trying to improve the agreement between modeled and measured variables is not practicable (on top of not being very scientific) due to the high-dimensionality of the parameter space and general uncertainties in the model.

- A direct instantaneous force measurement might contain artifacts due to other moving body parts. At least the muscles and halteres are known to move during flight, but also abdomen and head are not rigidly connected to the thorax. These organs have much larger masses than the wings. The model only accounts for aerodynamic and inertial effects of the wings.

Some of these issues are not easy to solve with the current setup. Nonetheless, these are the first attempts to directly compare modeled forces derived from a dynamically scaled robotic wing and instantaneous forces on the tethered insect. The similarity of the forces remains remarkable.

## 5.2 Lift control

Lift represents an important aspect of low-level flight control since it corresponds to the main component of flight force and must be continuously adjusted during flight maneuvers to compensate gravity. In this section, we apply a systems approach to reverse-engineering lift control.

A systems approach has the advantage of avoiding complexities arising from the low-level processes by treating the fly as a biological black box, where input-output relationships are described with a minimally-constrained model [Taylor et al., 2008].

Lift was chosen because it is one of the most suitable experimental paradigms for the tethered setup: Lift control is independent of the disruption of mechanosensor feedback in tethered experiments, because the fly, unlike airplanes or birds, does not rotate during lift adjustments (see the helicopter model by [Götz, 1964]).

Previous research on lift in fruit flies has mainly concentrated on the aerodynamic effects taking place to generate sufficient force. Lehmann et al [Lehmann and Dickinson, 1997] correlated the changes in wing beat frequency and amplitude to the mean lift force. In a further study, they looked at the ability of flies to vary wing beat frequency and amplitude as a function of lift [Lehmann, 2001]. Fewer studies have looked at the control aspects of lift. In locusts, Taylor [Taylor and Thomas, 2003, Taylor and Zbikowski, 2005] studied flight stability by analyzing the state space model of how the locust would have flown had it been instantly released from its tether.

A rigorous approach inspired from aeronautical engineering was used by Tanaka et al. [Tanaka and Kawachi, 2006] to study bumblebee flight control. Patterns were oscillated vertically at various frequencies and the responses were plotted in a Bode diagram. The frequency response was compared to frequency responses of other lift controllers.

Here, we extend this approach to a more detailed analysis of the fruit fly by employing the developed technologies.

### 5.2.1 Methods

Fruit flies were tethered to a MEMS micro force sensors (see Section 4.1) and placed under a high speed camera (see Section 4.2) in a visual flight simulator (see Section 4.3). The flies' reaction to vertically-oscillating patterns were recorded by measuring their lift forces and their kinematics . The oscillation frequency of the patterns was varied throughout the experiments.

**Visual patterns**

Two types of patterns were used in the experiments. The first consisted of a sinus grating that oscillated vertically: the "greenscale" intensity $g$ of a LED located at $x, y$ at time $t$ is described as:

$$g(x,y,t) = \frac{C}{2}(sin(2\pi \text{SF}(y + \frac{A_{in}}{2}sin(2\pi \text{TF}t))) + 1) \quad (5.1)$$

where $C$ is the contrast amplitude, SF is the spatial frequency [pixel$^{-1}$], TF is the temporal frequency [Hz] and $A_{in}$ is the oscillation amplitude [pixel].

The second visual pattern consisted of a sinusoid grating that underwent a step change in velocity.

5.2. LIFT CONTROL                                                                           61

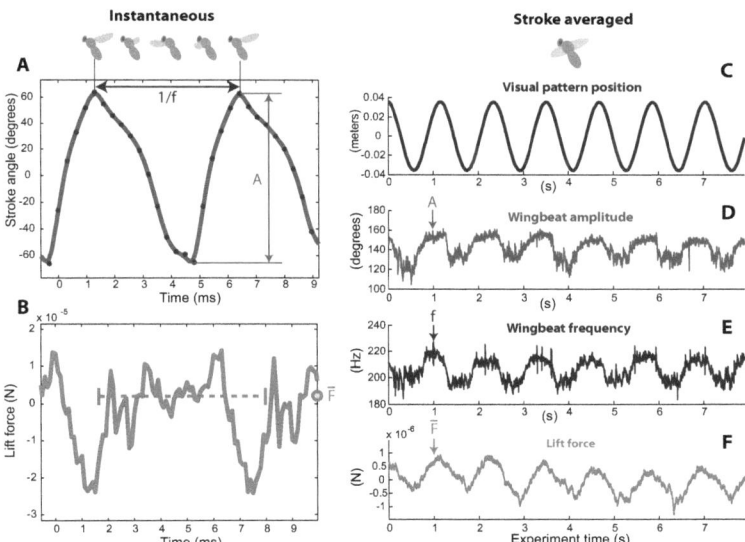

**Fig. 5.2:** Experimental setup and raw data. **A**: Wing kinematics extracted at 7000 fps. **B:** Lift forces measured by MEMS. **C:** Visual pattern position. The pattern is oscillated up and down (here at 0.8 Hz) **D:** Wing beat amplitude, extracted from instantaneous measurements. **E:** Wing beat frequency extracted from instantaneous measurements. **F:** Averaged lift forces. The forces were averaged for visualization purpose to remove the cyclic stroke components. Note the three orders of magnitude of time scales difference between the instantaneous (left) and stroke averaged (right) plots.

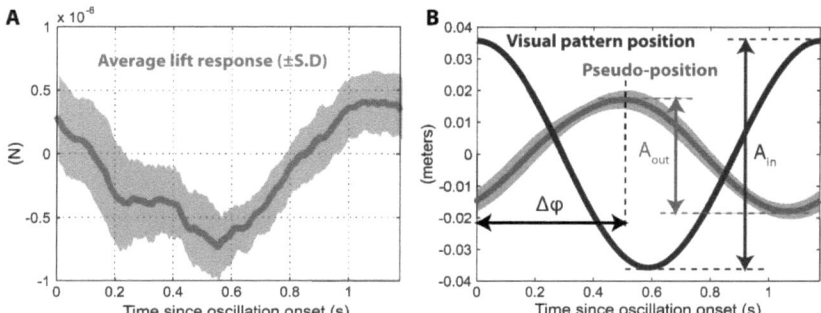

**Fig. 5.3:** Average response to an oscillation (in this figure, TF = 0.8 Hz, n=25 oscillations). **A:** Segmented force data. The force response to each visual pattern oscillation is overlayed. **B:** Input and output positions. Each individual force responses is numerically integrated to estimate the virtual positions of the fly. The virtual position of the fly is then put in relation to the visual pattern position by comparing the amplitudes $A_{in}, A_{out}$ and phase difference $\Delta\varphi$ (shown here as the phase difference between maxima)

$$g(x,y,t) = \frac{C}{2}\left(sin\left(2\pi \cdot \text{SF} \cdot y + \phi_{step}(t)\right) + 1\right) \quad (5.2)$$

$$\phi_{step}(t) = \begin{cases} 0 & \text{if } t < t_0 \\ 2\pi \text{TF}\,(t - t_0) & \text{if } t \geq t_0 \end{cases} \quad (5.3)$$

where $t_0$ is the step onset time. To calculate the position time course of a pattern, one must follow how a given phase in the grating moves through time.

Sinus gratings were chosen because they represent a single spatial frequency and therefore represent the most basic input to the visual system.

### 5.2.2 Experiments

We first characterized how the fly responded to vertically-oscillating patterns. For each measurement, the visual pattern properties, as defined in Equ.5.1, were randomly chosen in the following parameter space:

## 5.2. LIFT CONTROL

$$\begin{aligned} TF &\in [0.1, 7.0] & \text{Hz} \\ A_{in} &\in [20, 75] & \text{mm} \\ SF &= 21 & \text{m}^{-1} \\ C &= 7 & \end{aligned} \quad (5.4)$$

The visual pattern was presented to the fly for an interval of 20 s while the kinematics and forces were recorded. A typical measurement sample is shown in Fig.5.2.E-H. The average force response extracted from Fig.5.2.F-H is shown in Fig.5.5.A. To facilitate the extraction of a controller, the force response was then converted to the pseudo-position of a free-flying fly. The position corresponds to the altitude of a free flying fly producing the same force pattern. This conversion is used so that the input and output units are the same. To calculate the pseudo-position, the force is numerically integrated in an inertial newtonian model:

$$z = \frac{1}{m} \iint F(t) d^2 t \quad (5.5)$$

where $z$ is the pseudo-position, $m$ is the mass of the fly kg and $F$ is the resultant force measured at the sensor tip ($F = F_{fly} - m \cdot g$). A similar methodology was employed by [Tanaka and Kawachi, 2006] and this overall strategy is discussed in section 5.2.4. Fig.5.5.B shows the resulting pseudo-position pattern in comparison with the input stripe position.

### 5.2.3 Results

Given a sinusoidal stimulation, we expect the fly to respond with a periodic force trace. Indeed the response is approximately a sine of the same frequency as the input sine (Fig.5.2.E-H and 5.5.A). The double integration from Equ.5.5 causes the sine response to become even more clear (Fig.5.5.B). The amplitudes $A_{in}, A_{out}$ and phases $\varphi_{in}, \varphi_{out}$ are therefore sufficient to fully characterize the response.

The gain $G = 20 \cdot \log(A_{out}/A_{in})$ and phase difference $\Delta\varphi = \varphi_{out} - \varphi_{in}$ responses measured at different TF frequencies are then plotted in a Bode diagram (Fig.5.4). The gain response decreases from 40 dB at low frequencies with a slope of about forty decibels per decade. This is typical for a controller with a second order denominator. The variance of the gain is very low, which is remarkable for biological measurements.

The phase difference starts at $-100$ degrees and shows a linear decrease to ca. $-270$ degrees at 42 rad/s. This decay is expected for a system with a fixed time lag. The variance increases with frequency. To interpret the measured phase lags, we must consider the fact that insects react to pattern velocities, not position. In a

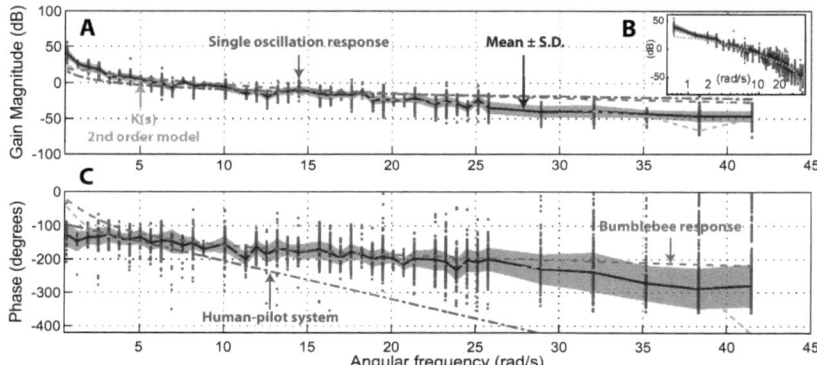

**Fig. 5.4:** Bode diagram of the fly's lift controller (input: visual pattern position, output: pseudo-position of fly - see Fig.5.5). **A:** amplitude gain (Lin-log). **B:** amplitude gain (Log-log) **C:** phase difference. Each red point represents a single oscillation. Data for ca. 8000 visual pattern oscillations measured on $n = 6$ flies. Blue line shows directional mean ± directional S.D.. Orange line shows the extracted model $K(s)$. Green – – line shows bumblebee response (from [Tanaka and Kawachi, 2006]) and purple – · – line shows human pilot response (from [McRuer and Jex, 1967])

## 5.2. LIFT CONTROL

sinusoidal stimulation, the velocity peak precedes the position peak by 90 degrees. Secondly, the flies reaction is a change in force output, which we integrated twice to get a position. The double integration (Equ.5.5) causes a $-180$ phase shift (compare Fig.5.5.A and 5.5.B). The ca. $-100$ degree phase difference measured at low frequencies in our position input/output diagram therefore corresponds to a 10 degree phase lag between velocity input and force output. The choice of input and output dimensions is discussed in detail below.

We used a the Frequency Domain Identification Toolbox available for Matlab (http://elecwww.vub.ac.be/fdident) to find an appropriate model for the data. We considered models with one to four zeros, one to four poles and with or without a time delay. We fit the 32 resulting models to the Bode measurements using the toolbox. About a half-dozen models had small cost functions. We selected the following one, because it maintained a small cost function even though it had among the lowest orders:

$$K(s) = \frac{5.24 \cdot 10^{-5} s^2 - 2.74 \cdot 10^{-5} s + 0.078}{0.0021 s^2 + 0.00658 s + 0.00513} \cdot e^{-0.03s} \tag{5.6}$$

This model has two zeros, two poles and a 30 ms time delay. The model fits well for the majority of the measurements, but diverges at both frequency extremes (see Fig.??). The good fit strengthens the hypothesis that the 40 dB/decade decay in the fly's lift response can be modeled as a second order denominator in the transfer function. The 30 ms time delay is in the same order of magnitude as visual delays measured in houseflies [Borst, 1986].

We also complement our frequency measurements with a set of velocity step response measurements (Equ.5.2). The flies reacts with step increase in force, which, once converted to position, can be compared with the estimation of the extracted model (see Fig.5.5) This provides a rapid method to verify that the extracted model $K(s)$ is valid for a different type of input.

### 5.2.4 Discussion

In this chapter, we present a novel approach to reverse-engineering lift response in fruit flies. To this end, we employed state-of-the-art microrobotic tools to characterize the flight dynamics of lift responses. We analyzed these measurements in a control theory framework, where the fly was considered as a biological black box, with known inputs and measurable outputs. We presented the Bode diagram of the fly's lift controller and compare it to lift controllers of other species and other man-made devices.

**Analyzing the fly as a controller**

By analyzing the fly from a control system's perspective, we benefit from several aspects. The analytical framework that has been set up for man-made controllers is readily available. The field of system identification helps us identify models and parameters from the raw data. The extracted system can easily be characterized in terms of stability and robustness. Such results can directly be compared with measurements performed with other species or robots, therefore making it easier to transfer knowledge in between the disparate fields of robotics and biology.

To benefit from these advantages, we had to quantify the input and output with the same units by deriving the pseudo-position of the fly from the force pattern measured with the MEMS. This was done to put the measurements in the framework of a controller, where a desired goal is fed in and the achieved output is measured. Such systems typically work in a feedback loop, where the desired and achieved are compared to generate an error signal for the controller. In our case, we are studying the open-loop characteristics of this system.

The systems approach also contains certain pitfalls. The first is to become too reductionist and forget that we are dealing with a biological entity. The biological complexity underlying the response in Fig.5.4 has much more to it than the extracted model $K(s)$ would seem to contain. The sensorimotor pathway is highly-interlinked and it is unlikely that the visual and motor control can be easily separated.

Another pitfall lies in the way we analyze the fly's controller. For instance, the use of the pattern's metric feature size in the Bode diagram is dangerous, because flies do not perceive metric dimensions but angular ones. In a wider flight arena, a pattern of the same metric size would be perceived at a smaller angle. In other words, the offset of the gain magnitude in Fig.5.4 is flight-arena-dependent. In a larger flight arena, the gains would be offset to the bottom.

Interestingly, this fact does not invalidate our gain magnitude data, because the fly (and any other vision-based flyer) must also cope with this difference in the real world: the way it sees (angular velocity) is fundamentally different from the spatial environment it is moving in. Can insects infer the distances of objects? Humans rely heavily on stereo vision and the recognition of objects of known sizes. Flies, however, have very little stereo overlap and are unlikely to be able to recognize complex patterns due to their coarse spatial resolution. Flies could rely on spatial cues, such as parallax, but in the absence of such cues, it is possible that the fly makes an assumption of distances. Our experimental setup allows to measure this assumption, by assuming that the fly, at low stimulation frequencies, generates the

## 5.2. LIFT CONTROL

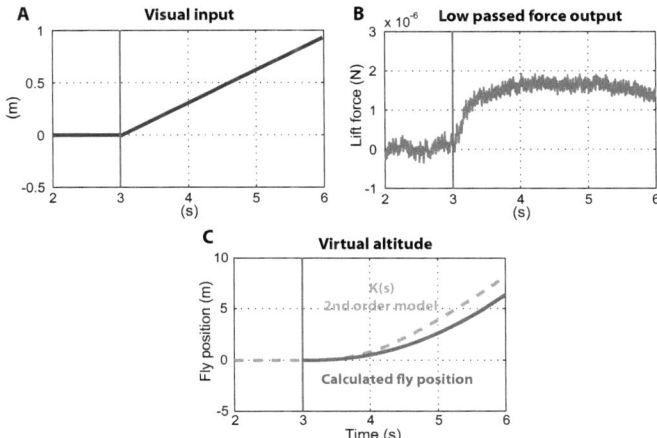

**Fig. 5.5:** Lift response to a step increase in pattern velocity. **A:** After three seconds, the pattern starts moving upwards at a constant velocity. **B:** The fly reacts by adjusting its mean force. The forces were low-passed with a zero-phase filter to remove the wing beat oscillations. **C:** By integrating the force, the fly's position is calculated and compared to the model $K(s)$ extracted from the Bode diagram.

exact force to compensate for the movements it is seeing.

Given the knowledge that flies react primarily to pattern velocities, one could ask why the input/output dimensions were not velocity. We chose position as an input/output dimension to facilitate the comparison with frequency response plots measured previously. Also, the Bode diagram for a velocity input/output (or any other time derivative of position) would be the same as Fig.5.4 as long as the measured response remained sinusoidal, because magnitude gains and phase differences are preserved in the differentiation of sinusoids.

### Comparison with other lift controllers

We can compare the extracted frequency response to frequency responses measured on other flying systems. As discussed above, the dependency of the gain magnitude on the setup size limits how much comparison can be made directly. However, the fitted models contain interesting similarities that are independent of setup size. For human controlled aerial-vehicles, the measurements pointed to a general first order system $\frac{A}{s} \cdot e^{-\tau}$ [McRuer and Jex, 1967]. In bumblebees, a second order model fitted the measurements best (however, measurements were sparse [Tanaka and Kawachi, 2006]). These systems all had time delays that went from 200 ms for the bumblebee to over a second for humans. Smaller flying organisms face inherently less stable flight, because the instantaneous forces they generate become larger compared to their body masses. Similarly to highly maneuverable (and thus instable) fighter jets that require the help of rapid low-level controllers for humans to pilot them, small flying organisms are likely to have developed faster controllers to cope with their instable dynamics.

### 5.2.5 Conclusion

These experiments bring up a whole new set of relevant questions. For instance, we have assumed a linear time-invariant controller in the fruit fly. It would be interesting to measure the response to a superposition of two input (temporal) frequencies, to verify the superposition property expected from a LTI system. Closed-loop experiments could confirm the controller extracted here, and tackle the difficult question of feedback within the biological system itself. Finally, the extraction of spatial dimensions from optic flow is highly relevant to any vision based flying object. A broader analysis of step responses to different pattern sizes and speeds will help characterize this.

In summary, we have presented a rigorous approach to reverse-engineering microaerial flight control in biology. This approach generalizes the biological system so that it is directly comparable to artificial entities. Furthermore, this work shows how engineering can provide both strong technical tools and an analytical framework for the understanding of biological sensorimotor pathways, a task that is highly relevant to the design of artificial devices at similar dynamic scales.

## 5.3 Thrust control

The control of thrust in fruit flies is an important modality, because the fly spends most of its time flying longitudinally. We also decided to analyze thrust because a complementary experiment in the free flight setup, where flies fly in a wind tunnel, analyzes longitudinal flight control. The reproduction of a similar experiment in two different setups allows to compare the detailed measurements of the tethered setup

## 5.3. THRUST CONTROL

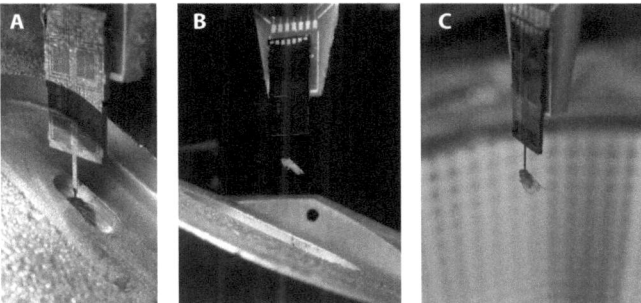

**Fig. 5.6:** Thrust control experiments **A:** Tethering **B:** Tethered fly **C:** Fly and arena.

to the naturalistic conditions in the free fly flight setup. In particular, the effect of the tether can be analyzed.

### 5.3.1 Methods

The experiments consisted in using the three-axis force sensor (see Section 4.1 and ChapterB) to concurrently measure lift, thrust and pitch torque (see Fig.5.6). The fly was placed in the flight arena (see Section 4.3). The visual pattern shown to the fly consisted in vertical sine gratings. The sine gratings were initially immobile, but after a few seconds, they underwent a step increase in horizontal velocity, similarly to what was described for the step experiments in the lift paradigm (see Equ.5.2). We varied the spatial frequency of the sine gratings, as well as the temporal frequency of the velocity step.

### 5.3.2 Results

The flies typically responded with a step increase in lift, thrust and pitching torque (see Fig.5.7). We therefore used the mean to quantify the responses to different visual parameters. Since we wanted to compare these results with the results in the wind tunnel, we divided the mean force response by the mass of the fly, and therefore obtained the acceleration the fly would have experienced had it not been tethered [1]. The resulting "pseudo-acceleration" is then color-plotted against TF and SF (see

---
[1]This calculation is based on a purely inertial model of the fly, and therefore neglects drag.

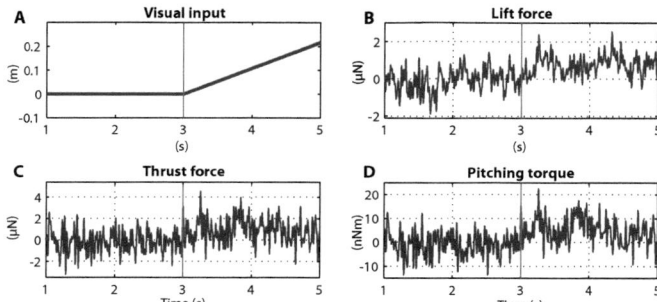

**Fig. 5.7:** Typical low-passed thrust response data ($f_c = 45$ Hz) **A:** Stripe pattern position **B:** Lift force response **C:** Thrust force response **D:** Pitching torque response.

Fig.5.8).

### 5.3.3 Discussion and conclusion

The data measured in the tethered setup does not give as clean of results as what was measured in free flight [Rohrseitz and Fry, ]. We believe the main cause of this is that the tether disrupts the normal pitch response of the fly, and we therefore do not get a very naturalistic response: without any feedback that it has pitched forward, the fly may be acting erratically to achieve this first goal.

Another potential cause lies in the way the fly was visually stimulated. For technical reasons, the experiments were all performed on the circular arena, with half the pixels showing the left eye pattern and half the pixels showing the right eye pattern. This does not represent very well what the fly sees in naturalistic conditions: there is no distortion of elements far away and the expansion/contraction created in front and behind the fly might cause an escape/fixation response instead of the intended speed adjustment response. To overcome this problem, the easiest is to perform the experiments using the planar panels or even better, mounting the tethered fly inside the wind tunnel.

In conclusion, these experiments are more of a proof-of-principle of the thrust paradigm in tethered flight. The experiments can be extended to analyze the biomechanics of thrust: by measuring in the wind tunnel, the fly's response to varying wing speeds can be analyzed in detail and compared with the fly's visual response.

## 5.3. THRUST CONTROL 71

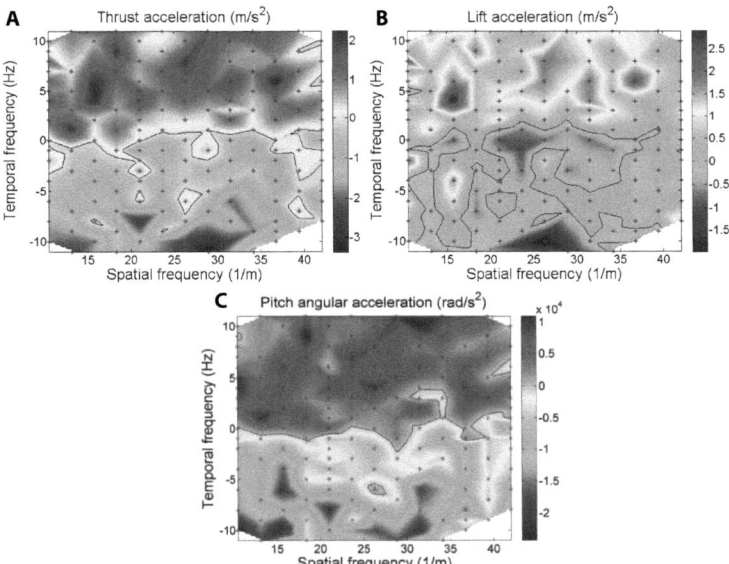

**Fig. 5.8:** Overview of the thrust experiments. **A:** Thrust pseudo-acceleration ($\Delta F_{thrust}/m_{fly}$ where $\Delta F_{thrust}$ is the mean difference in thrust force before and after the step occurs.) **B:** Lift pseudo-acceleration **C:** Pitch pseudo-acceleration. The "*" represent the points in the parameter space that were actually measured. The points in between "*" were interpolated. Definitions of SF and TF as in Equ.5.2.

Goetz's influential helicopter model [Götz, 1964] can also be put under scrutiny with these methods. Our preliminary results seem to go against his theory, however, more experiments are needed to make an unequivocal statement.

# Chapter 6

# Closing remarks

Each chapter in this thesis has ended with a specific discussion and conclusion that is not necessary to repeat. Instead, I will summarize the overall contributions of the work and take a look at the future.

This thesis has taken place in the exciting interaction between biology and robotics. A set of novel technical tools were developed: a specifically-designed MEMS sensor, fabricated by IRIS, was calibrated and implemented in the experimental setup to measure the instantaneous flight forces of tethered fruit flies in up to three concurrent directions. A high speed vision system was developed to extract, in real time, the kinematics of both wings. A LED panel system was fabricated to generate appropriate visual stimuli.

These tools were applied in a reverse-engineering approach to identify low-level control strategies in fruit flies. A first experiment looked at the biomechanics of tethered flight, and compared the instantaneous force measurements with the predictions of a model fed with the fly's measured kinematics. The second experiment used a systems approach and analyzed lift control. A bode diagram of the fly's lift control was extracted and compared to other species. In the third experiment, the fly's longitudinal speed control was analyzed by measuring the lift, thrust and pitching torque produced by the fly in response to a visual stimulation in forward speed. In a side project, the fly's controller was used to drive a robot in closed-loop, allowing us to explore the coupling characteristics of both.

## 6.1 Contributions

Through the development and application of new engineering technologies to a biological problem, this thesis has made significant contributions to both fields.

For engineering, the contributions lie in the extension of measurement techniques into previously unreachable spatiotemporal scales: smaller, faster forces for the MEMS, high image velocities for the DWBA and rapid control of visual stimuli for the flight simulator. Through the application of these technologies to a scientific problem, they went through the stringent questioning phase of the scientific method: each possible measurement artefact had to be considered and ruled out before a hypothesis could be validated or not. The amount of effort this represents compared to initial proof-of-principle experiments (pure technology with very little science) should not be underestimated. Unlike what the quite linear "development-application" structure of this thesis seems to point to, there is no such thing as a development phase followed by an application phase. Both are intertwined. The technologies have gone through numerous iterations to reach a "product" level. At each iteration, biological measurements were performed and questions were asked that led to new implementation rules.

For biology, the benefits lie in the possibilities offered by the technology. It is often so that technical innovations do not just lead to better measurements for an existing paradigm, but rather that the possibilities that were unleashed create a whole new way of thinking that replaces the old one [Kuhn, 1962]. Very often, technical innovations lie at the heart of new biological discoveries that were not even searched for in the first place. One of these paradigms that is emerging here is the concept of a *systems approach*. In such an approach, the fly's internal complexities are bundled up into a black box model. The analysis is thus simplified as it allows to concentrate on the sensory inputs and behavioral outputs of the fly, ignoring for the most part the details of what is going on in between. Through specific experimental paradigm (choice of stimulation, use of genetic tools), a particular part of the black box can be identified, thus preventing the risk of being too general in the analysis. Another experimental paradigm rendered possible, or even imaginable, through the technologies developed in this thesis is the analysis of mechanosensors in tethered flight: the real-time capabilities of the DWBA are put to use to generate wing-phase-locked forces to the halteres, thus creating pseudo-Coriolis forces that simulate a body rotation.

## 6.2 Outlook

It is likely that the interest in small flying organisms will increase in coming years, as the technologies to analyze them and to build similarly-scaled artificial flyers becomes available. The contributions of this thesis will therefore serve future work in understanding important aspects of insect flight.

More pragmatically, genetics is likely to expand the horizon of possibilities. The ability of genetic tools to isolate a particular process in the sensorimotor pathway combined with the technologies and the the rigorous, quantitative, systems approach shown here, will be very powerful.

# Appendix A

# MEMS calibration

## A.1 Static calibration of three-axis sensors

As mentioned in 4.1.5, the calibration of the three-axis sensors involves using a single-axis reference sensor and measuring the force/force plots in different directions.

### A.1.1 Calibration procedure

The two sensors are brought in contact under a microscope (see Fig.A.1). The two electrical readouts are synchronized so that the forces can be measured concurrently on each (see Fig.A.2).

Given that there are no reference 3DOF sensors at the range and resolution used here, the calibration is performed using a single axis reference sensor and the procedure is repeated in three configurations (see Fig.A.3). In each of the three calibration positions, static measurements of 5000 data points are recorded. Between the measurements, the force applied is increased stepwise until it reaches the maximum value of 500 µN for the 3DOF sensor in x direction or 600 µN for the 3DOF sensor in y direction, and is then decreased again.

The data from the four capacity readouts (one is not used for the 3DOF sensor) is averaged and collected with the data from the other increments (see Fig.A.4 for an example of calibration data).

### A.1.2 Extraction of calibration matrix

The further calculation is based on Simon Muntwyler's Master thesis [Muntwyler, 2006]. The slopes of the first order least square fits in the plots in Fig.A.4 are actually

78                                      APPENDIX A.  MEMS CALIBRATION

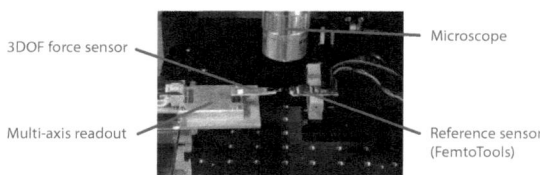

**Fig. A.1:** 3DOF calibration setup: the 3DOF sensor is brought in contact with a reference 1DOF sensor. Forces on each are recorded simultaneously. The operation is repeated in different directions.

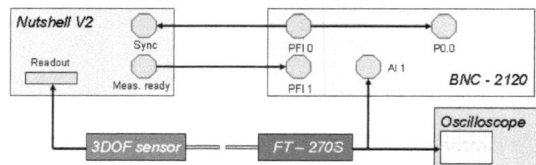

**Fig. A.2:** Connection diagram for the multi-axis calibration.

**Fig. A.3:** The three calibration steps: A) x calibration B) y calibration C) yy calibration.

## A.2. DYNAMIC CONSIDERATIONS IN DISPLACEMENT/FORCE SENSORS

the entries of the matrix in Equ.A.1. C2, C3 and C4 are the measured differential capacities, while Rx, Ry and Ryy represent the force values measured by the reference sensor in the three calibration positions. The MEMS sensor characteristic curve is supposed to be linear and the slopes of C2, C3 and C4 are approximated with first order fits. Fig.A.4 shows that this is a good assumption for reference forces up to 300 µN.

$$\begin{pmatrix} C2 \\ C3 \\ C4 \end{pmatrix} = \begin{pmatrix} \frac{dC2}{dRx} & \frac{dC2}{dRy} & \frac{dC2}{dRyy} \\ \frac{dC3}{dRx} & \frac{dC3}{dRy} & \frac{dC3}{dRyy} \\ \frac{dC4}{dRx} & \frac{dC4}{dRy} & \frac{dC4}{dRyy} \end{pmatrix} \cdot \begin{pmatrix} Rx \\ Ry \\ Ryy \end{pmatrix} \quad (A.1)$$

Equ.A.1 is then numerically inverted. Equ.A.2 shows the values of the inverted matrix as they were calculated for sensor C.

$$\begin{pmatrix} Rx \\ Ry \\ Ryy \end{pmatrix} = \begin{pmatrix} 0.0018 & -0.0093 & -0.0397 \\ -0.0267 & 0.0838 & 0.0004 \\ 0.1013 & -0.0927 & 0.0007 \end{pmatrix} \cdot \begin{pmatrix} C2 \\ C3 \\ C4 \end{pmatrix} \quad (A.2)$$

Now, to convert the values Rx, Ry and Ryy to the actual forces and torques acting on the sensor, a simple matrix multiplication is performed:

$$\begin{pmatrix} Fx \\ Fy \\ Mz \end{pmatrix} = \begin{pmatrix} 1 & 0 & 0 \\ 0 & 1 & 1 \\ 0 & 0 & 2 \end{pmatrix} \cdot \begin{pmatrix} Rx \\ Ry \\ Ryy \end{pmatrix} \quad (A.3)$$

Finally, combining Equ.A.2 and Equ.A.3 leads to the 3x3 calibration matrix shown in equation A.4 with the values calculated for sensor C.

$$\begin{pmatrix} Fx \\ Fy \\ Mz \end{pmatrix} = \begin{pmatrix} 0.0018 & -0.0093 & -0.0397 \\ 0.0747 & -0.0089 & 0.0011 \\ 0.2027 & -0.1853 & 0.0013 \end{pmatrix} \cdot \begin{pmatrix} C2 \\ C3 \\ C4 \end{pmatrix} \quad (A.4)$$

## A.2 Dynamic considerations in displacement/force sensors

### A.2.1 Introduction

As mentioned in 4.1.5, the MEMS sensor measures forces indirectly: A force causes a deflection of the sensor probe. For static and low-frequent signals, this relationship

80  APPENDIX A. MEMS CALIBRATION

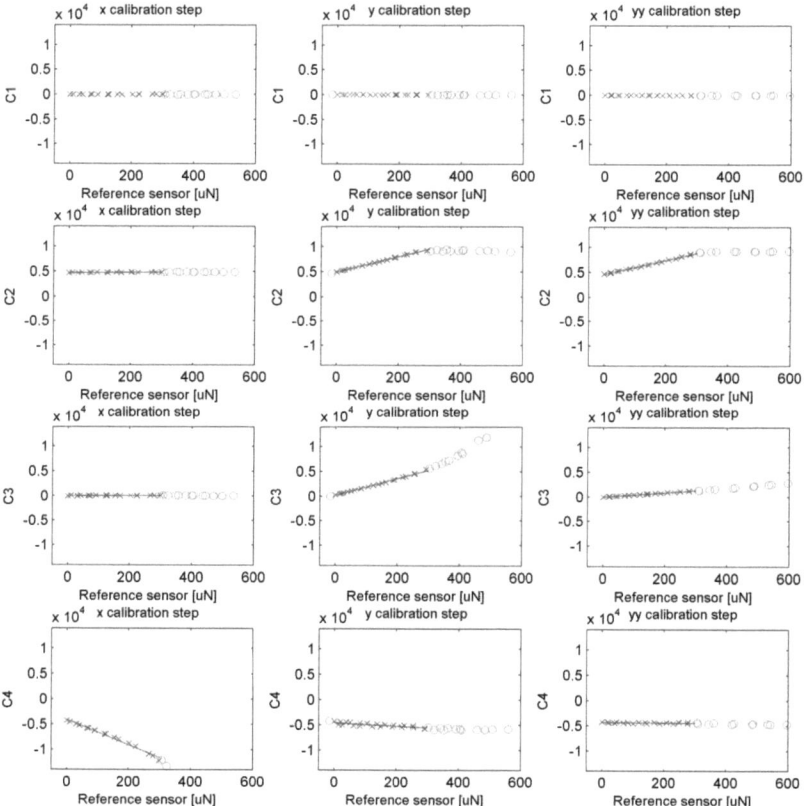

**Fig. A.4:** Data from a full calibration. The data points marked 'x' are used in the calibration matrix extraction. The data marked 'O' are disregarded because they either are above the calibration range, or their standard deviation was too high.

## A.2. DYNAMIC CONSIDERATIONS IN DISPLACEMENT/FORCE SENSORS 81

behaves proportionally. However, this proportionality gets lost at high frequencies where dynamic effects take place.

**Sensor characteristics**

The sensors use a differential triplate comb drive configuration, as seen in Fig.A.5. The two stationary comb drives were electrically isolated by placing the capacitor plates on each side of the moveable comb set.

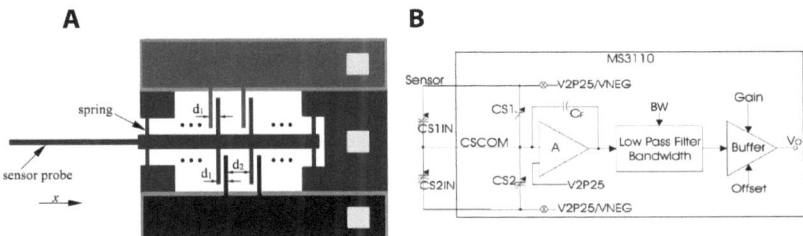

**Fig. A.5:** A: Schematic illustration of the micro force sensor. The two stationary comb drives are situated on each side of the moveable comb set (adapted from [Sun et al., 2005]). B: MS3110 MEMS capacitive readout circuit. A change in the capacity causes a change in the output voltage (adapted from www.irvine-sensors.com/pdf/MS3110DatasheetUSE.pdf).

Because of the differential configuration of the capacitive pair, the relationship between deflection and output voltage is quasi-linear. The force applied to the probe causes a deflection which changes the electrical capacitance of both capacities and finally the output voltage of the MS3110 Universal Capacitive Readout IC (Irvine Sensors Corporation). The functional block diagram of this readout chip is illustrated in Fig.A.5.B.

**The limits of static calibration**

At low frequencies, force and displacement can be assumed to be proportional. Since most users of these force sensors are interested in mostly static forces, FemtoTools performs a static calibration, where force and voltage are proportional. The S540 sensors have a gain of about 80 µN/V.

When measuring rapidly-varying forces, such as the forces produced by the wing beat of a fruit fly, force and deflection are not proportional anymore. This happens as the frequency content of the input forces approaches the eigenfrequency of the MEMS' mechanical structure. The oscillations of the MEMS' movable structure cause amplitude and phase shifts between the input force and the resulting deflections. Given that the resonant frequency of the sensor without any extra mass added to the probe is around 2 kHz, we expect significant effects of the vibrations to already appear after several hundred Hz , especially in the presence of an added mass such as the fly which brings the resonance frequency lower. This limit is quantified at the end of the section, see Fig.A.13.

The difference between the static and a dynamic calibration can be seen in Table 4.2. The electronic readout transfer behavior is defined by the MS3110 MEMS readout chip of Irvine Sensors Corporation (www.irvine-sensors.com) and is assumed to be constant for all frequencies and sensors.

The goal of this chapter is to first gain sufficient theoretical knowledge and then experimentally identify $f^{-1}$, the inverted frequency response of the sensor that will allow us to reconstruct the original force pattern that generated the recorded voltage signal.

### A.2.2 Theoretical Modeling

The goal of this section is to identify a proper model of the force sensor so that we can calculate an estimate $\widehat{F}(t)$ of the original forces $F(t)$ from the measured voltage pattern $V(t)$ (see Fig.A.6).

To determine this model, we split the system "sensor" up into two subsystems: the mechanical and the electrical subsystems. The mechanical system $G$ models how external forces displace the comb structure. The electrical system $h(x)$ models how a displacement of the comb structure affects the output voltage. To reconstruct the original force from the measured voltage, we must invert these two models.

#### Model of the mechanical system

As we can see in Fig.A.5.A, the moveable part of the sensor is fixed with four springs. To characterize the behavior of this structure, we started with the commonly used mass-spring-damper model. As we will show in the following section, this model turned out to fit very well our experimental measurements and so we never had to search for alternative models.

## A.2. DYNAMIC CONSIDERATIONS IN DISPLACEMENT/FORCE SENSORS83

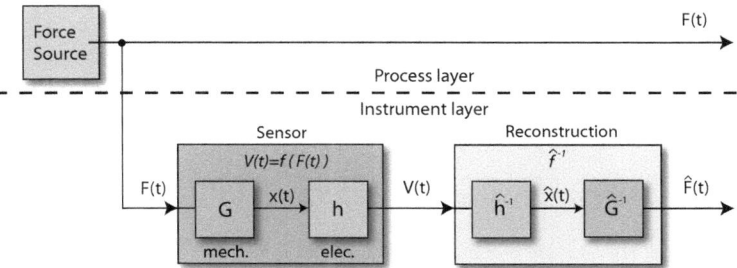

**Fig. A.6:** Overview of the dynamic calibration process. The force source (upper left) produces a force $F(t)$ that generates a displacement $x(t)$ on the force sensor. The relationship is characterized by the mechanical model $G$. The displacement further causes a change in capacitance which is picked up by the electrical readout and generates a voltage $V$. This transduction is characterized by the function h(x). The modeling effort here tries to identify the inverse models $\widehat{h}^{-1}$ and $\widehat{G}^{-1}$ that allow to reconstruct the force $\widehat{F}(t)$ from the measured voltage $V$. The "^" denote "estimated"' values.

Using Newton's second law of motion, we can write the ordinary differential equation of a mass-spring-damper system as:

$$m\ddot{x} + b\dot{x} + kx = F(t) \tag{A.5}$$

where k is the spring constant (N/m ), b is the damping constant (Ns/m ) and m is the mass of the sensor (kg ). To derive the transfer function between the force $F(t)$ as input and the deflection $x(t)$ as output, we take the Laplace transform of Equ.A.5. This yields:

$$m(s^2 X(s) - sx(0) - \cdot x(0)) + b \cdot (s \cdot X(s) - x(0)) + k \cdot X(s) = F(s)$$

Taking $x(0) = \dot{x}(0) = 0$ as initial condition yields:

$$ms^2 X(s) + bsX(s) + kX(s) = F(s) \tag{A.6}$$

Which leads to:

$$G(s) = \frac{X(s)}{F(s)} = \frac{1/m}{s^2 + 2\delta\omega_0 s + \omega_0^2} \tag{A.7}$$

where $\delta = b/(2\sqrt{k \cdot m})$ is the damping coefficient and $\omega_0 = \sqrt{k/m}$ is the eigenfrequency. This transfer function can then be inverted in the reconstruction to obtain a force $F$ from a deflection $x$: $\widehat{F}(s) = G^{-1}(s)\widehat{X}(s)$.

## Model of the electronic readout system

The force applied to the force sensor causes a deflection of the capacitors, which induces an output voltage of the readout circuit. The characteristics of this conversion is described in the datasheet of the MS3110 Universal capacitive readout IC by Irvine Sensors Corporation (www.irvine-sensors.com):

$$V(t) = \text{Gain} \cdot V_{2P25} \cdot 1.14 \cdot \frac{C_{S2}(t) - C_{S1}(t)}{C_F} + V_{ref} \qquad (A.8)$$

where $V(t)$ is the output Voltage and $V_{ref}$ is the reference voltage. Gain, $V_{2P25}$ are constants and $C_F$ is selected to optimize the input sense capacitance range. $C_{S2}$ and $C_{S1}$ are the MEMS capacitor pair and are defined as:

$$C_{S2} = n \cdot \epsilon \cdot \frac{A}{d_1 - x} + n \cdot \epsilon \cdot \frac{A}{d_2 + x} \qquad C_{S1} = n \cdot \epsilon \cdot \frac{A}{d_1 + x} + n \cdot \epsilon \cdot \frac{A}{d_2 - x} \qquad (A.9)$$

where n is the number of capacitive comb drives, $\epsilon$ the permittivity of air and A,d1,d2 the geometric specification of the sensor as shown in Fig.A.5.A. The following table summarizes all constants and parameters:

**Table A.1:** Parameters of the transfer equation of the electronic readout system and the values that are used for them in the Equ.A.10 for both sensor types.

| Parameter | FT-S270 | FT-S540 |
|---|---|---|
| Area $A$ | $2.5 \cdot 10^{-8}$ m$^2$ | $2.5 \cdot 10^{-8}$ m$^2$ |
| $n$ | 50 | 50 |
| $d_1$ | 5 µm | 5 µm |
| $d_2$ | 20 µm | 20 µm |
| $\epsilon$ | $8.859 \cdot 10^{-12}$ F/m | $8.859 \cdot 10^{-12}$ F/m |
| $C_F$ | 4 pF | 4 pF |
| Gain | 2 V | 4 V |
| $V_{2P25}$ | 2.25 VDC | 2.25 VDC |
| $V_{ref}$ | 2.25 V | 2.25 V |

Therefore the final transfer function of the MS3110 electronic readout system is defined as:

$$V(t) = h\left(x(t)\right) = \underbrace{\frac{\text{Gain} \cdot V_{2P25} \cdot 1.14 \cdot 2 \cdot n \cdot \epsilon \cdot A \cdot (d_2^2 - d_1^2)}{C_F \cdot d_1^2 \cdot d_2^2}}_{K} \cdot x(t) + V_{ref} \qquad (A.10)$$

## A.2. DYNAMIC CONSIDERATIONS IN DISPLACEMENT/FORCE SENSORS

Again, to receive the reconstruction transfer function, we take the inverse of function A.10[1], $\widehat{x}(t) = h^{-1}(V(t)) = \frac{V(t) - V_{ref}}{K}$

**The complete transfer function**

To calculate the overall voltage-to-force transfer function, we can simply combine the two transfer functions from above:

$$\widehat{F}(s) = f^{-1}(V(s)) = G^{-1}(s) \cdot h^{-1}(V) \quad (A.11)$$

$$= m \cdot (s^2 + 2\delta\omega_0 s + \omega_0^2) \cdot \left(\frac{V - V_{ref}}{K}\right) \quad (A.12)$$

To get to the transfer behavior of the system in the time domain, we have to apply an inverse Laplace transform of the Equ.A.11. A separation of the equation yields:

$$\widehat{F}(s) = \frac{m}{K} \cdot \left(s^2 \cdot (V - V_{ref}) + 2\delta\omega_0 s \cdot (V - V_{ref}) + \omega_0^2 (V - V_{ref})\right) \quad (A.13)$$

Further, using the properties of Laplace-transformation yields:

$$\widehat{F}(t) = \widehat{f}^{-1}(V(t)) \quad (A.14)$$

$$= \frac{m}{K} \cdot \left(\ddot{V}(t) + 2 \cdot \delta \cdot \omega_0 \cdot \dot{V}(t) + \omega_0^2 \cdot (V(t) - V_{ref})\right) \quad (A.15)$$

Once the model has been verified and its parameters identified, this function can be used to implement an off-line filter that reconstructs an estimate $\widehat{F}(t)$ of the original forces $F(t)$ from the voltage measurements $V(t)$.

### A.2.3 Experimental identification

In Equ.A.14, certain model parameters are still unknown: the mass $m$, the damping coefficient $\delta$ and the eigenfrequency $\omega_0$. Ideally, to identify these parameters, one would use a sinusoidal force source and increase its frequency while keeping its amplitude constant. The frequency response of the sensor could then directly be measured, and the three parameters above could be extracted.

---

[1] Since this function is only a gain and offset, it does not play a role whether we use it in time domain or frequency domain.

Unfortunately, there are no ideal force sources like the one described above, so we had to find another technique to extract the missing parameters.

The simplest technique is to use the sensor's combs as electrostatic actuators: a sinusoidal voltage is applied between the combs, which generates an electrostatic force in between them. The resulting displacement of the sensor probe is precisely measured with a laser vibrometer. A major drawback of this technique is that the exact voltage/force relationship is unknown. Depending on this relationship, the expected frequency response occurs sometimes at twice the frequency than the excitation. All in all, it is difficult to know exactly how the sensor is being excited. [2]

We decided to implement another experimental paradigm to have a clear understanding of the system. The technique consisted in mounting the entire sensor on a piezo transducer (see Fig.A.7). The piezo was aligned so that the movement it induced on the force sensor could be measured by the laser vibrometer. In this experiment, both the sensor's body and it's movable part are oscillating, so two lasers are necessary to measure the relative movement of the two.

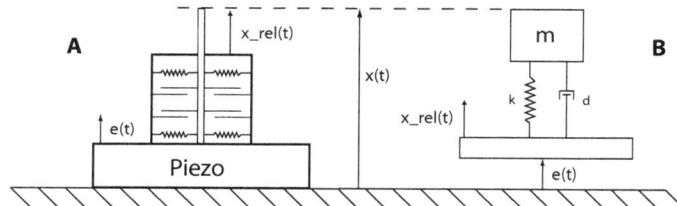

**Fig. A.7:** A: Dynamic calibration setup schematic. The piezo generates a displacement $e(t)$ of the entire sensor. This generates further oscillations of the sensor movable structure $x(t)$. The relative displacement between the sensor's base and its movable part is denoted $x_{rel}(t)$ B: Assumed mass-spring-damper system for the experimental setup with the Piezo excitation under the scanning vibrometer.

The mass-damper-model of this system yields:

$$m\ddot{x} = -b \cdot \dot{x}_{rel} - k \cdot x_{rel} \tag{A.16}$$

where $x_{rel} = x - e$ is the relative deflection between the sensor and the moveable part and $e(t)$ is the displacement initiated by the piezo, as shown in Fig.A.7.

---

[2] Such a confusion has probably led FemtoTools to under-estimate the resonance of their sensors by a factor two.

## A.2. DYNAMIC CONSIDERATIONS IN DISPLACEMENT/FORCE SENSORS 87

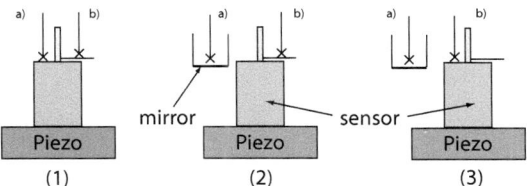

**Fig. A.8:** Illustration of the measurements done on the laser vibrometer. The measurements are enumerated as in the text. a) is the reference laser. b) is the measurement laser.

Solving this equation in the Laplace domain:

$$G_2(s) = \frac{X(s)}{E(s)} = \frac{2\delta\omega_0 s + \omega_0^2}{s^2 + 2\delta\omega_0 s + \omega_0^2} \tag{A.17}$$

This is not exactly the same transfer function than the one we are trying to identify from the sensor: $G(s)$, see Equ.A.7. However, the parameters are the same, so we can use $G_2(s)$ to identify $G(s)$.

**Frequency response experiments**

Using the experimental setup described in the previous section, several Bode diagrams of the system were measured. The measurements were done in three steps (see Fig.A.8).

1. The reference laser is aimed on the sensor's body. The measurement laser on a small wing that is rigidly connected to the sensor's probe.

2. The reference laser is covered with a small cup with a mirror inside while the measurement laser is pointed on the movable part of the sensor. In this setup, the absolute deflection of the probe is measured.

3. The reference laser is still covered with a mirror-cup, but the measurement laser is pointed on the sensor's body. Using the same excitation signal for the piezo as in experiment 2, the absolute deflection of the [3] of the sensor's body is measured.

---

[3]This part moves in this experimental setup, but is fixed when measuring forces

APPENDIX A. MEMS CALIBRATION

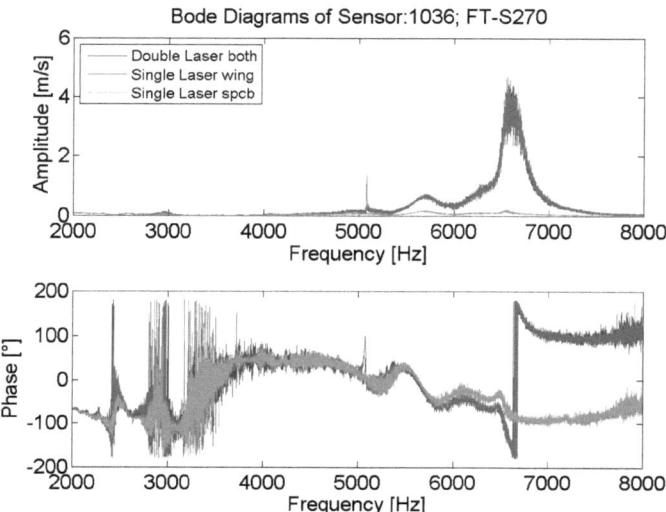

**Fig. A.9:** Measurements of the deflection velocities of the FT-S270 sensor (Nr. 1036) in frequency domain. The diagram is almost the same for the measurements with the double laser (blue), and the single laser pointed onto the moveable part (green). The red curve is the body part of the sensor and shows a much lower amplitude (spcb means laser pointing to the body)

Theoretically, only the first experiment is necessary. However, the quality of the data obtained with the two laser experiment was inferior to the experiments 2 and 3. We therefore mainly used experiments 2 and 3. Fig.A.9 shows the result for the FT-S270 sensor. The vibrometer we used in these experiments measures only the velocities of the deflections. For this reason the data where saved in m/s and not in m what is negligible in our case. Using the equation A.17 we can see that the transfer function remains the same for position and velocity, input and output dimensions.

For the FT-S270, it is easy to identify the resonance frequency, whereas for the Ft-S540 sensor it is more difficult, due to a second peak around 4.9 kHz and several lower peaks in the whole frequency spectrum. These peaks may appear from different modal forms, e.g from different direction of deflection while oscillating with the piezo. The 4.9 kHz peak, in particular, shows up clearly in both sensors, and can therefore

## A.2. DYNAMIC CONSIDERATIONS IN DISPLACEMENT/FORCE SENSORS

be assumed to be independent of the sensor's springs. It is likely caused by the piezo or the connection between the piezo and the sensor, and is disregarded in the following analysis.

Based on these measurements, the resonance frequencies of the peaks are found to be $\approx 6.6$ kHz for the FT-S270 and $\approx 2.57$ kHz for the FT-S540 sensor. We will now discuss how we can extract the model parameters from the Bode diagram.

### Parameter identification

The parameters that we want to identify from the transfer function $G_2$ (Equ.A.17) and the measured Bode diagram (see Fig.??) are: the mass of the system $m$, the damping constant $b$ and the spring constant $k$.

**Mass of the sensor $m$**   The easiest way of obtaining the mass of the sensor is simply to weight it. We removed the movable sensor part of an unused sensor and put it on a micro-scale. Due to the small size of the sensor, 5 capacitor plates broke off. Since their weight is small with respect to the rest of the probe, they were neglected. The final weight was found to be at about $m = 0.7 \pm 0.4$ μkg [4]. This value of the mass is assumed to be constant for all sensors of both types.

**Spring constant $k$**   If we had the Bode diagram of the sensor system (Equ.A.7), the spring constant would simply be the steady state gain (gain at $s = 0$). Unfortunately, the system we measured does not have this property. Its steady state gain is 1. We therefore have to adopt another approach. There are two possibilities to estimate the spring constant:

- Experimentally measuring the spring constant by adding a constant force by another force sensor to the sensor with zero frequency. In this case we know both the force and the displacement of the sensor, so we can calculate $k = F/x$

- Using the resonance frequency $\omega_{res} \approx \omega_0 = \sqrt{k/m}$ (for low damping) from the Bode diagram.

The second way is easier and less time consuming, because there is no need to setup another experiment.

---

[4] This is about the same as the weight of a fruit fly ($\approx 1$ mg).

**Damping constant $b$**  It is more complex to determine the damping constant $b$, or further the damping coefficient $\delta$. Several ways are proposed:

- Using the bandwidth of the system, respectively the bandwidth of the measured Bode diagram. This method requires a clear phase response of the system.

- Trying to derive the numeric transfer function of the system using the System Identification Toolbox of MATLAB and then compare this transfer function with the model A.17 to identify the parameters.

- If the eigenfrequency and mass is already defined (see identification of spring constant k), we can simply derive $\delta$ and furthermore $b$ of the system by combining $b = 2\delta\sqrt{k \cdot m}$ with the following relationship:

  If $\delta < 1/\sqrt{2}$, then: (from [Guzella, 2007])

$$\mid G_2 \mid_{(\omega=\omega_0)} = 20\log_{10}(\frac{1}{2\delta\sqrt{1-\delta^2}}) \tag{A.18}$$

The vibrometer measurements give us no clear information of the phase response of the system and no clear amplitude peak, so the first method doesn't seem to be a good way. The second way is not very consistent either because we have no proof of correctness of the model and only a small frequency band of the whole spectrum. Given that the frequency response of the system provides us a clear peak, the third method is the best to estimate the damping constant.

**Table A.2:** Mechanical parameters of the sensors derived from the measurements

| Sensor | FT-S270 | FT-S540 |
|---|---|---|
| Mass $m$ | 0.7   $mg$ | 0.7   $mg$ |
| Eigenfrequency $\omega_0$ | 41'928   rad/s | 16148   rad/s |
| Eigenfrequency $f_0$ | 6673   Hz | 2570   Hz |
| Spring constant $k$ | 1230.6   kg/s$^2$ | 182.5257   kg/s$^2$ |
| $\delta$ | 0.0146 | 0.053 |
| Damping constant $b$ | $8.564 \cdot 10^{-4}$ kg/s | $11.99 \cdot 10^{-4}$ kg/s |

With these Parameters defined, we obtain the following numerical values for the transfer function A.17 of the mechanical model:

$$\textbf{FT-S270:} \quad G_{270}(s) = \frac{1224.3s + 17.579 \cdot 10^9}{s^2 + 1224.3s + 17.579 \cdot 10^9} \tag{A.19}$$

## A.2. DYNAMIC CONSIDERATIONS IN DISPLACEMENT/FORCE SENSORS

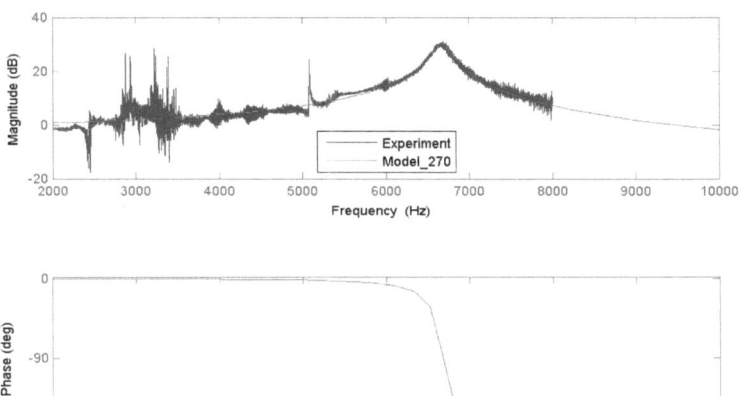

**Fig. A.10:** Bode diagrams of the measured sensor with the modeled system, using the parameters collected by the experiments for the Sensor FT-S270.

$$\text{FT-S540:} \quad G_{540}(s) = \frac{1711.688s + 260.76 \cdot 10^6}{s^2 + 1711.688s + 260.76 \cdot 10^6} \quad \text{(A.20)}$$

Fig.A.10 shows the Bode diagrams of the modeled sensor. The experimental models derived with the numerical transfer functions A.19 and A.20 show a remarkable fit to the main resonance response of the measured system. This proves our assumption of the mass-spring-damper model. Note that we did not show the phase information because it could only be measured very poorly with the technique applied here on the laser vibrometer.

### A.2.4 Dynamic calibration implementation

This section deals with the more practical aspects of the dynamic calibration: how do we choose the model parameters for a new sensor and how do we implement a digital filter to transform the voltages back into forces?.

**System parameter choice**

Some of the model parameters are the same for all sensors. For instance, the electrical readout parameters can be considered constant (see TableA.1). The mechanical model parameters are all a little more sensitive to the microfabrication steps, and are therefore prone to variations from one sensor to another. However, the planar dimensions are the most sensitive to microfabrication parameters. In other words, of the three mechanical model parameters, the mass and the damping coefficient will not be very sensitive because they are hardly affected by changes in the planar dimensions. The spring constant, on the other hand, has a cubic dependency on the width of the springs (designed to be 5 µm wide). Therefore, the spring constant needs to be adjusted for each sensor. The technique we used in this section to identify the spring constant was to look at the resonant frequency in the frequency response of the force sensor. While this technique is precise, it is quite time-consuming to perform the laser vibrometer experiments for each sensor. Luckily, the spring constant can be inferred from the much simpler static calibration (already performed by FemtoTools). By combining Equ.A.8 and Table4.2, we obtain:

$$k = k_1 \cdot K \quad (A.21)$$

where $k$ is the spring constant N/m , $k_1$ is the static calibration constant N/V *and* $K$ is the cumulative gain of the electrical readout V/m (see Equ.A.8).

The mass of the probe and the damping coefficients can be assumed to be constant (see Fig.A.11). The mass of the tethered object (in our case, the fruit fly) will reduce the resonant frequency and must therefore be taken into account: it is added to the mass of the probe.

**Digital filter implementation**

To transform the voltage series $V(t)$ into a force estimate series $\widehat{F}(t)$, we must apply $f^{-1}$ (Equ.A.14). The approach we used consisted in numerically differentiating the measurement. Numerical differentiation is very sensitive to noise amplification and we need a second derivative of our signal. To minimize this noise, we used a method called "symmetrical differential quotient":

## A.2. DYNAMIC CONSIDERATIONS IN DISPLACEMENT/FORCE SENSORS

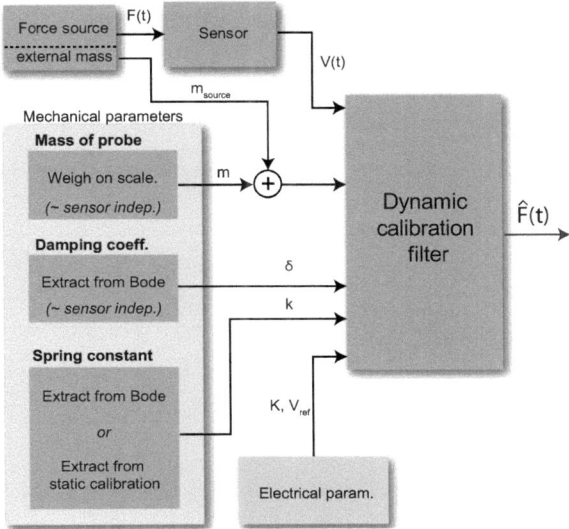

**Fig. A.11:** Overview of parameter choice and signal flow in the dynamic calibration implementation.

Combining the definition of the Fourier rows:

$$f(x+h) = f(x) + \frac{h}{1!}f'(x) + \frac{h^2}{2!}f''(x) + \ldots + \frac{h^k}{k!}f^k(x) \tag{A.22}$$

$$f(x-h) = f(x) - \frac{h}{1!}f'(x) + \frac{h^2}{2!}f''(x) - \ldots \tag{A.23}$$

and neglecting terms of higher order yields the symmetric differential quotients for first and second order:

$$\triangle_h^1(x) = \frac{f(x+h) - f(x-h)}{2h} = f'(x) + \frac{h^2}{6}f'''(x) + O(h^2) \tag{A.24}$$

$$\triangle_h^2(x) = \frac{f(x+h) - 2\cdot f(x) + f(x-h)}{h^2} = f''(x) + O(h^2) \tag{A.25}$$

where the step size $h$ is the sampling time $t = 1/f_s = 1\cdot 10^{-4}$ in our case.

## A.2.5 Validation of the dynamic calibration

For signals with low frequencies ($s \approx 0$), we have a redundant calibration: the dynamic calibration should match the static calibration. This allows us to verify that the steady state gain of the dynamic calibration is correct.

**Table A.3:** Sensitivities of both model and Femto tools fact sheet and their error for both considered sensors.

| Sensor Nr. | Type | Sensitivity Model | Sensitivity Fact sheet | Error |
|---|---|---|---|---|
| 1036 | FT-S270 | 1155.27 µN/V | 1050.8 µN/V | 9% |
| 1044 | FT-S540 | 85.68 µN/V | 73.2 µN/V | 14.6% |

These errors are in a very acceptable range and have different potential causes:

- Error in the mass of the sensor probe, either due to the broken capacitor plates or the resolution of the scale

- Error in the measurement of the eigenfrequency of the system (this would result in a different damping and spring constant)

- The transfer function of the electronic readout system ?? was approximated to be linear.

## A.2.6 Results

To test the dynamic calibration of the sensors, the implemented filters have been used with data obtained with different fruit flies tethered to the probe. The flies were visually stimulated with light sources at different frequencies. Since the FT-S540 sensors are much more sensitive, most measurements have been done using this sensor instead of the FT-S270 sensors.

The dynamic calibration filter has been tested on several flies using the FT-S540 sensors. The dynamic calibration effect is very visible: both the amplitude and the phase of the signal is corrected (see A.12).

## A.2.7 As of when is a dynamic calibration necessary?

We started off this chapter by saying that force sources containing high frequency information would require a dynamic calibration. We are now able to quantitatively determine as of which frequency the static calibration does not suffice. In Fig.A.13,

## A.2. DYNAMIC CONSIDERATIONS IN DISPLACEMENT/FORCE SENSORS

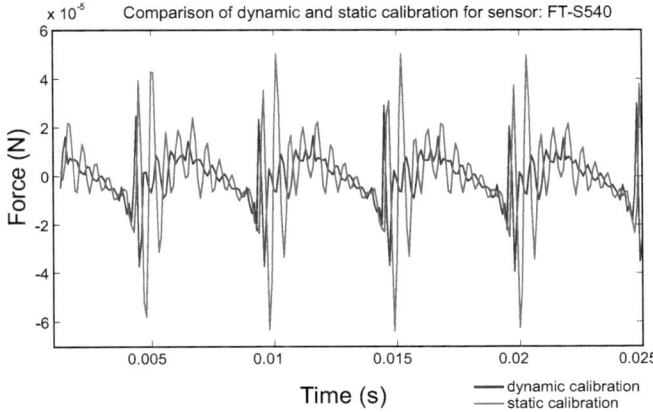

**Fig. A.12:** Comparison of the static calibration with the dynamic calibration. The phase and amplitude shift caused by the resonance effects of the sensor are very visible.

we show the calibration gain as a function of frequency for both static and dynamic calibrations. The static calibration is independent of frequency. The difference in between the two curves can be interpreted as the error of force amplitude made by employing a static calibration on a time-dependent signal. The point where this error exceeds 3dB is marked with a dotted line. This happens above $\approx$ 1000 Hz for the FT-S540 Sensor and above $\approx$ 2200 Hz for the FT-S270. Note that in this case, the mass of the fly was taken into account ($\approx$ 1 mg. For a heavier object, these thresholds would be lower.

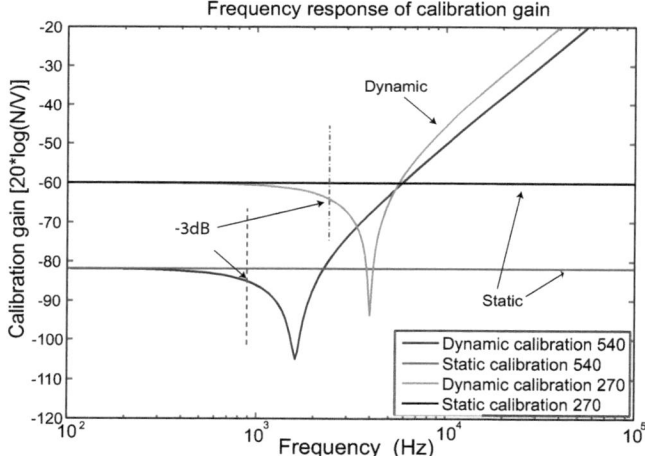

**Fig. A.13:** Calibration gain N/V as a function of frequency (mass of fly included).

# Appendix B

# Implementation of a quasi-steady model of lift generation

This chapter covers the model of lift generation that was compared to instantaneous forces and kinematics measured on the fruit fly (see Section 5.1).

## B.1 Model variables and constants

To describe stroke kinematics, the three instantaneous stroke angles $\alpha$, $\theta$, $\phi$ were defined similar to the standard literature, as illustrated in fig. ?? [Sane and Dickinson, 2001]:

$\alpha$   aerodynamic angle of attack
$\theta$   vertical stroke-deviation
$\phi$   horizontal stroke-position

The aerodynamic angle of attack $\alpha$ is different from the physiological angle of attack. It is defined as the angle between the leading edge of the wing $\vec{x}_{le}$ and the direction of motion of the wing tip $\vec{U}_t$ (both unitary vectors):

$$\cos\alpha = \vec{x}_{le} \cdot \vec{U}_t$$

Apart from these variables, all modeled forces contain constants deriving from wing morphology, physiology and fluid properties. The values were gathered from Ellington [Ellington, 1984b], Lehmann [Lehmann and Dickinson, 1997] and Fry [Fry et al., 2005] (see TableB.1).

| variable | value | unit | description |
|---|---|---|---|
| $\rho_a$ | 1.2 | $kg/m^3$ | density of air [Lehmann and Dickinson, 1997] |
| $\rho_w$ | 1200 | $kg/m^3$ | density of wing material (solid cuticle) [Ellington, 1984b, Lehmanr |
| $R_w$ | 2.47E-3 | $m$ | wing length [Lehmann and Dickinson, 1997] |
| $\bar{c}_w$ | 0.80E-3 | $m$ | mean wing chord [Ellington, 1984b, Lehmann and Dickinson, 199 |
| $h_w$ | 1.33E-6 | $m$ | wing thickness [Ellington, 1984b, Lehmann and Dickinson, 1997] |
| $S$ | 1.98E-6 | $m^2$ | wing surface $= R_w \bar{c}_w$ |
| $m_w$ | 3.16E-9 | $kg$ | wing mass $= \rho_w R_w \bar{c}_w h_w$ [Ellington, 1984b, Lehmann and Dickinse |
| $x_{cm}$ | 0.7 | $R_w$ | relative length-wise centre of mass position [Fry et al., 2005] |

**Table B.1:** Constants defining wing morphology, physiology and fluid properties, as they were used in the inertial and aerodynamic model.

## B.2 Inertial forces

According to standard mechanics, the inertial force that a flapping wing exerts on the fly's body corresponds to

$$\vec{F}_{acc} = -m_w \ddot{\vec{x}}_{cm}$$

where $m_w$ is the wing mass and $\ddot{\vec{x}}_{cm}$ is the position of the centre of mass of the wing.

The length-wise position of the wing's centre of mass is situated approximately at 70% of the line from base to tip ($\hat{d}_x = 0.7$) [Fry et al., 2005], the chordwise position $\hat{d}_y$ was kept as a variable within the model at $0.2 - 0.5$ in units of average chord length measured from the leading edge (no measured value found in the literature).

## B.3 Aerodynamic forces

The instantaneous forces generated by a flapping wing are represented as the sum of three separate force components: added mass, translational and rotational force. Each force component is assumed to act normal to the wing surface [Sane and Dickinson, 2002]. Therefore, only the magnitude[1] of the different forces has to be summed to obtain the combined force:

$$F_{aero} = F_{admass} + F_{trans} + F_{rot}$$

---
[1]Different signs have to be taken into account.

## B.3. AERODYNAMIC FORCES

- **Translational force** The translational force depends on the instantaneous wing tip velocity magnitude $U_t$ and the aerodynamic angle of attack $\alpha$. As customary the translational force is divided into a drag part (acting anti-parallel to the direction of motion) and lift part (acting perpendicular to the direction of motion). Drag and lift coefficients were determined empirically [Dickinson et al., 1999][2] and summed through vector addition:

$$U_t^2 = R^2(\dot{\theta}^2 + \dot{\phi}^2)$$
$$C_l(\alpha) = -0.004 + 1.92\sin(1.95\alpha + 0.04)$$
$$C_d(\alpha) = 2.10 + 1.57\sin(2.04\alpha - 1.64)$$
$$F_{trans} = \frac{1}{2}\rho_a S U_t^2 \hat{r}_2^2 \sqrt{C_l^2(\alpha) + C_d^2(\alpha)}.$$

- **Rotational force** Using Kutta-Jukowski aerodynamic theory, rotational force has been estimated theoretically [Sane and Dickinson, 2002].

    It depends on the wing tip velocity $U_t$ and the first derivative of the aerodynamic angle of attack $\omega = \dot{\alpha}$. Further, it depends on the axis of $\alpha$-rotation of the wing determined by the non-dimensional axis of rotation $\hat{x}_0$: a rotation about the leading edge corresponds to $\hat{x}_0 = 0$, about the trailing edge to $\hat{x}_0 = 1$.

    The rotational coefficient $C_r$ and corresponding force are given by:

$$C_{rot} = \pi(0.75 - \hat{x}_0)$$
$$F_{rot} = C_r \rho U_t \omega \bar{c}_w^2 R \int_0^1 \hat{r}\hat{c}_w^2(\hat{r})d\hat{r}$$

- **Added mass force** The added mass component has been estimated theoretically [Sedov, 1965] to correspond to

$$F_{admass} = \rho\frac{\pi}{4}R^2\bar{c}^2(\ddot{\phi}\sin\alpha + \dot{\phi}\dot{\alpha}\cos\alpha)\int_0^1 \hat{r}\hat{c}^2(\hat{r})d\hat{r}$$

(The second summand in formula (3) in [Sane and Dickinson, 2002] has been eliminated since W.B. Dickson confirmed [personal communication] that it does not give accurate and physically plausible results).

---

[2]modified values from personal communication with W.B. Dickson

## APPENDIX B. QSS MODEL OF LIFT GENERATION

- **Morphological integrals** The formulae for the revised quasi-steady model contain two integrals comprising the non-dimensional wing chord and non-dimensional radial position (for definitions see [Ellington, 1984b]). They depend on wing morphology and can be estimated assuming an elliptic wing shape. Instead, I used values that W.B. Dickson kindly disclosed [personal communication]:

| integral | value |
|---|---|
| $\int_0^1 \hat{r}\hat{c}^2(\hat{r})d\hat{r}$ | 0.829 |
| $\int_0^1 \hat{r}^2\hat{c}(\hat{r})d\hat{r} = \hat{r}_2^2$ [Ellington, 1984b] | 0.372 |

# Appendix C
# Exploration of bio/robot coupling

In this chapter, we present a non-invasive 'Cyborg' system as a biorobotic platform to explore the emergent behaviors resulting from the coupling of a tethered fly and a wheeled robot (see Fig.C.1).

The idea to implement a Cyborg Fly came from Vasco Medici and Steven Fry when we were asked to create a workshop for roboticists and biologists interested in flying objects. The goal was to foster discussions on the interactions of these two fields.

While robotics can benefit from biological research, robotic platforms can conversely enhance biological understanding by serving as testbeds to explore and validate hypothesis of sensorimotor pathways [Franceschini et al., 2007, Halloy et al., 2007]. The freedom in the design of the robotic device allows to create arbitrary experimental situations that can be used in a repeatable way to explore "cases" that would be unpractical or impossible to test on the organism itself. Such an approach may also lead to the identification of emergent behaviors that would be difficult to identify from the observation of the organism's behavior alone [Abbott, 2007].

Though such biorobotic implementations are intriguing, the transfer between a target biological system and its robotic counterpart is non-trivial and prone to misconceptions [Webb, 2006, Datteri and Tamburrini, 2007]. The transfer starts by modeling the biological process. The choice of the model's complexity level is crucial, because a too general model will lose meaningfulness and a too complex model may be impossible to realize in an artificial system.

The model is then implemented in a robotic system. Because biological systems have different building blocks and operate at different spatiotemporal scales than engineered devices, this important step always involves a certain *level of abstraction*. One of the most important and difficult effort is to properly take into account the dif-

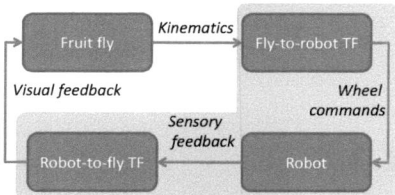

**Fig. C.1:** Fly-in-the-loop: the fruit fly's normal sensory feedback loop is replaced with a non-invasive biorobotic coupling (grey box).

ferences in physical implementation. For instance, the underlying functional principle of an insect's compound eye might be successfully reproduced in an artificial system using very different building blocks (e.g. a CCD pixel instead of an ommatidium), as long as the extracted optic flow is the same.

Another big difference is the spatiotemporal scaling: biological organisms are often orders of magnitude smaller and faster than their artificial counterparts. Even if we were able to recreate an exact but scaled copy of a biological process, the optimality of the biological system might still be lost through the scaling.

Finally, once the implementation has taken place, we verify that the artificial system does indeed reproduce the target biological behavior. The comparison can be made using different criteria, i.e. this step involves another level of abstraction. A too lenient criteria might fail to notice strong drawbacks of the implemented system.

In conclusion, the process of coupling biological and robotic systems contains several pitfalls. There is the risk that the biological model is taken out of context and becomes therefore meaningless. The success of a biologically inspired robot to reproduce the behavior of the target biological system is not sufficient to make conclusions about the validity of the couplings that have been used in between. These misconceptions may lead to false 'optimality' claims for biomimetic robots or even to misleading scientific claims in the case of robotic-augmented biology.

To investigate this biology/robotic coupling without making assumptions, one approach is to actually take the biological control system (the insect), and let the organism control the artificial system itself. The organism becomes part of the control loop, in a special kind of non-invasive "cyborg" system (See Fig.C.1). A recent example of such a system was presented by Hertz *et al*, where a cockroach drove a mobile robot by walking on a modified computer mouse ball [Hertz, 2008].

The relevance of this work can be separated into three aspects that address the biorobotic issues stated above. First, the platform allows a direct interaction with the fly's sensorimotor pathway, giving a model-free paradigm to gain a functional understanding of the processes at play. Second, through the use of flexible transfer functions, the platform allows a vast exploration of the spatial and temporal couplings in between the robotic and biological system. Finally, the platform represents a clear artificially-closed-loop paradigm, with a visual, interpretable, output state - the robot's behavior - and therefore helps clarify this complex concept.

Through the understanding of biorobotic couplings, such cyborg systems may contribute to medicine, in cases where a prosthetic replaces a disabled biological subsystem while being controlled by the patient's brain. To this mean, researchers in the field of neuroprostethics have been studying techniques to have animals directly control robotic devices [Kositsky et al., 2003, Sato et al., 2008, Wessberg et al., 2000, Abbott, 2006].

## C.1 The Cyborg system

The concept of the Cyborg system is shown in Fig.C.2 and pictures of the system are represented in Fig.C.3. In the Cyborg Fly, the intended corrective flight maneuvers of a tethered fly (Fig.C.2.A) are measured with a high speed camera (C.2.B). These data are the input to a user-definable transfer function (C.2.C) that generates motor commands for a wheeled robot (C.2.D). As the robot moves through a cluttered environment, it returns sensory information (visual and range data) to a second transfer function (C.2.E) that generates an image for the flight arena (C.2.F). The fly responds to the visual stimulus, closing the loop. Each of these individual components are described in detail below.

### C.1.1 Fruit fly

We tethered wild-type fruit flies (*Drosophila melanogaster*) to a tungsten rod using standard procedures [Tammero and Dickinson, 2002].

### C.1.2 Digital wing beat analyzer

The digital wing beat analyzer (see Section 4.2) robustly extracts the wing position in real time at 7000 Hz (3500 Hz per wing) and with minimal latencies (maximum 150 µs). The system feeds this data into an extended Kalman filter. The Kalman

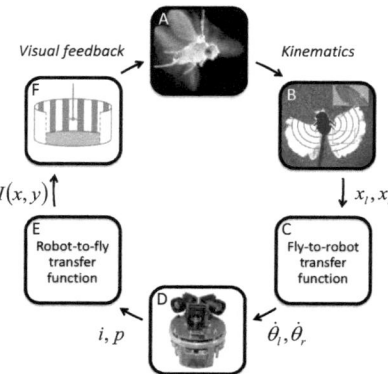

**Fig. C.2:** The Cyborg Fly is composed of six processes. Each process is described in detail below.

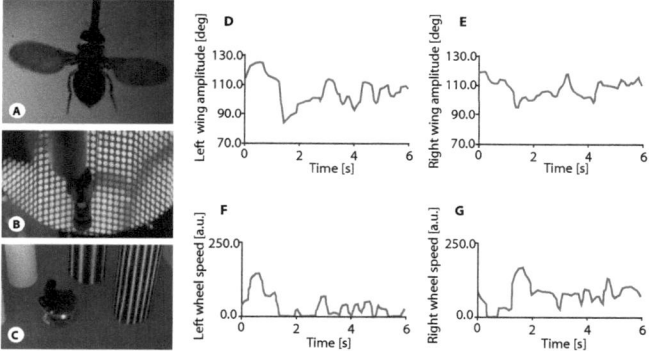

**Fig. C.3:** Experimental setup and example of data. A: tethered fruit fly viewed from high speed camera. B: Inside view of flight arena C: E-puck robot inside maze D: Left wing amplitude measured from high speed camera E: Right wing amplitude F: Left wheel speed G: Right wheel speed. The wheel speed commands were generated from the wing kinematics and fly-to-robot transfer function in Equ.C.3. Initially, the robot is turning right, but then a sudden decrease in the fly's left wing amplitude makes it turn left.

## C.1. THE CYBORG SYSTEM

filter's state vector $x$ provides a real-time estimation of wing beat frequency $f$, amplitude $A$, phase $\phi$ and mean angular position $m$, which can be transformed into motor commands to the robot.

### C.1.3 Fly-to-robot transfer function

The fly-to-robot transfer function transforms the four Kalman parameters extracted by the high speed vision system into velocity commands to the wheels of the robot as follows:

$$\begin{pmatrix} \dot{\theta}_l \\ \dot{\theta}_r \end{pmatrix} = \text{TF}_{\text{fly-to-robot}}(x_l, x_r) \tag{C.1}$$

$$= \text{TF}_{\text{fly-to-robot}}(A_l, A_r, m_l, m_r, \phi_l, \phi_r, f) \tag{C.2}$$

where $\dot{\theta}_l$ and $\dot{\theta}_r$ are the angular speed commands to the left and right wheels of the robot [°/s], $x = \begin{bmatrix} A & m & \phi & f \end{bmatrix}^T \in \mathbb{R}^4$ is the state vector of the left and right wing's Kalman filters as described in the previous paragraph. Because flies cannot separately modulate their left and right wing beat frequencies, $f_l = f_r = f$ in (C.2). Note that each term in (C.1) and (C.2) are intrinsically time dependent and (C.1) may be a function of state vectors at different discrete times ($x(t), x(t-1), ..., x(t-n)$).

The choice of (C.1) is completely unconstrained. Fruit flies, however, present a small repertoire of robust control responses: Yaw torque is controlled by varying the difference of stroke amplitude ($A_l - A_r$) [Fry et al., 2003]. Lift is controlled by increasing the mean speed of both wings ($A \cdot f$) [Lehmann and Dickinson, 1997]. Thrust is controlled by a modulation of both pitch (induced by changes in the mean stroke position $m$) and total force magnitude (proportional to $A \cdot f$) [Fry et al., 2005].

As a note of caution, tethered flight is subject to some quantifiable artifacts [Fry et al., 2005], because the tether disrupts the normal sensory feedback to the fly. In other words, a tethered fly's responses provide meaningful control signals, but one should be careful when generalizing these results to free flight.

Given this knowledge, a first logical approach is to use a control output that mimics the control of flight. Two general classes of transfer functions can be built that should result in a robot mimicking the fly's intended maneuvers:

- Yaw response of the robot is coupled to the yaw response of the fly:

$$\begin{pmatrix} \dot{\theta}_l \\ \dot{\theta}_r \end{pmatrix} = G_1 \begin{bmatrix} 1 & -1 \\ -1 & 1 \end{bmatrix} \begin{bmatrix} A_l \\ A_r \end{bmatrix} + H_1 \tag{C.3}$$

where $G_1$ [s$^{-1}$] is the gain from stroke amplitude difference (proportional to the fly's turning torque) to wheel velocity difference (robot wheel velocity), and $H_1$ is a set speed constant.

- Forward velocity of the robot is coupled to the lift/thrust response of the fly:

$$\begin{pmatrix} \dot{\theta}_l \\ \dot{\theta}_r \end{pmatrix} = G_2 \cdot f \cdot \begin{bmatrix} 1 & 1 \\ 1 & 1 \end{bmatrix} \begin{bmatrix} A_l \\ A_r \end{bmatrix} + H_2 \tag{C.4}$$

where $G_2$ is the unit-less gain from wing velocity (proportional to lift force) to forward robot velocity, and $H_2$ is an added speed constant.

## C.1.4 Robot

We implemented the experiments on an e-puck (www.e-puck.org) robot equipped with an array of three linear cameras (102 pixels) each and 8 proximity sensors. The control of the robot wheels was performed at 50 Hz while the readout of the linear camera and proximity sensory was performed at 10 and 20 Hz, respectively. All communications with the robot were performed through a Bluetooth wireless interface. The robot control is programmed using microcontroller-specific C code.

## C.1.5 Robot-to-fly transfer function

The visual image and/or the proximity sensor from the robot is employed to generate an image for the flight arena of the fly:

$$I(x, y) = \text{TF}_{\text{robot-to-fly}}(\boldsymbol{i}, \boldsymbol{p}) \tag{C.5}$$

where $I(x,y)$ represents the image to be shown on the flight arena. $\boldsymbol{i} = (i_1, i_2, ..., i_{306}) \in \mathbb{R}^{306}$ is the linear image from the three cameras of the robot and $\boldsymbol{p} = [p_1, p_2, ..., p_8] \in \mathbb{R}^8$ is the output of the eight proximity sensors.

Again, there are an infinite number of transfer functions that can be implemented. In this case, we focused on stimuli that are known to create robust yaw and lift responses in flies. Table C.1 gives a list of the ones we used.

## C.1.6 LED visual flight simulator

The LED visual flight simulator (see Section 4.3) was refreshed at 30-400Hz, depending on the choice of the robot-to-fly transfer function.

## C.1. THE CYBORG SYSTEM

Table C.1: Employed robot-to-fly transfer functions

| Robot-to-fly transfer function type | Description | Image |
|---|---|---|
| A. Undistorted robot view | The fly directly sees the output of the linear cameras. | |
| B. Nearest object defines azimuth and speed of expansion/contraction | The center of expansion is set in the same direction as the nearest object. The distance to the object defines the speed of expansion. A center of contraction is generated at the exact opposite of the expansion. | |
| C. Brightest object defines stripe position | The azimuth of the brightest object in the linear camera image defines the azimuth of the bright stripe in the flight arena. | |
| D. Object distance drives sinusoidal grating vertical speed | If the nearest object is to the right (left), the pattern will go up (down, respectively). The closer the object, the faster the pattern motion. | |

### C.1.7 Overall system control

The processes were controlled through a LabVIEW interface (National Instruments, www.ni.com) that provides a simple mean to manage the information flow and give the user the possibility to vary the transfer function parameters (see Fig.C.4). The high speed camera system ran on a separate computer and sent the Kalman state vector via UDP packets to the LabVIEW computer. The LabVIEW program read the UDP input and transformed it into motor commands by employing the user-defined fly-to-robot transfer function. In parallel, the LabVIEW program logged the sensory information from the robot and transformed it into a 2D image (via the user-defined robot-to-fly transfer function) that was sent to the CompactRIO device through Ethernet. The CompactRIO device transformed the image into I2C commands and sent them to the individual panels.

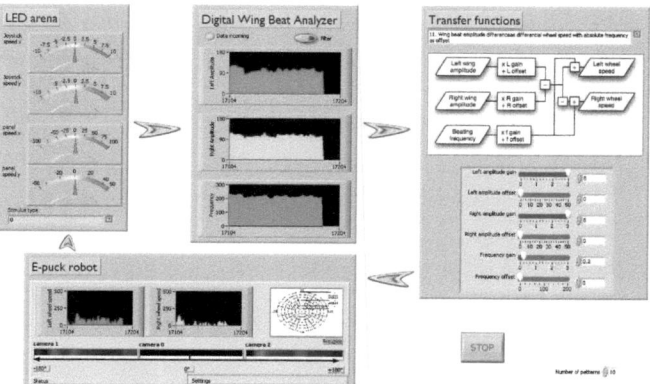

**Fig. C.4:** LabVIEW user interface: the user can visualize the fly's response (center) choose a transfer function type and set its gains (right side). The robot actuator and sensor data are logged at the bottom and transferred to the panel system via a second transfer function (left).

The hardware latency of the whole cyborg system is under 50 ms, mainly limited by the necessity to stream the robot's sensory information to the computer via Bluetooth.

## C.2 Experiments

### C.2.1 Naturalistic feedback

In a first set of experiments, we used the most direct, natural, form of feedback. The robot's movements were matched to the predicted free-flight movements of the fly. This was done by combining Eq. (C.3) and (C.4) to have the robot a) turn proportionally to the difference in wing beat amplitude and b) advance at a speed proportional to the mean wing speed. Experimental data is shown in Fig.C.3.E-H. The fly was shown a live view from the robot's cameras (Table C.1.A). These feedbacks were similar to the feedback used in the cockroach experiments by Hertz *et al* [Hertz, 2008].

This would seem naively to be the best choice of feedback, because it represents the most realistic mapping of the environment onto the animal, and vice versa. In-

## C.2. EXPERIMENTS

terestingly, the system did not perform robustly in these circumstances. The robot would repeatedly crash into obstacles, not showing the desired behavior.

### C.2.2 Amplified naturalistic feedback

To overcome these issues, we enhanced the visual feedback to the fly while keeping the same fly-to-robot transfer function: We generated expansion/contraction stimuli based on the mean distance of surrounding objects: as an object got closer, the expansion would accelerate (Table C.1.B). In this amplified situation, the Cyborg system showed robust obstacle avoidance behavior. After adjusting the transfer function gains, the robot drove through the arena for several minutes while avoiding walls and obstacles.

### C.2.3 Inverted response feedback

Flies are known to generate a robust stripe fixation behavior. We used this tracking behavior to generate obstacle avoidance in the robot. To this end, we placed the bright stripe pattern (TableC.1.C) in the opposite direction of the closest object. The tracking behavior of the fly was therefore coupled with the inverse behavior in the robot: obstacle avoidance.

### C.2.4 Decoupled response feedback

Taking this artificial coupling even further, we coupled completely unrealistic behaviors together. For instance, we used the lift response of the fly to control the turning of the robot:

$$\begin{pmatrix} \dot{\theta}_l \\ \dot{\theta}_r \end{pmatrix} = G_3 \cdot \begin{bmatrix} 1 & 1 \\ -1 & -1 \end{bmatrix} \begin{bmatrix} f \cdot A_l - H_3 \\ f \cdot A_r - H_4 \end{bmatrix} + H_5 \qquad (C.6)$$

To adjust the feedback accordingly, we used the mean distance of obstacles on each side of the robot to control the speed of ascent/descent of a stripe pattern (TableC.1.D). These experiments succeeded equally well in generating stable obstacle-avoidance behavior.

In summary, all of these experiments show that the emergent behavior of the system produces many non-intuitive results. Obviously, it is not so that a direct representation of the mapping leads to the expected, stable, behavior, because of the large differences in terms of temporal and spatial scaling. Secondly, if the robot does show similar behavior to that of a fly, it doesn't necessarily represent a truly biomimetic implementation.

## C.3 Discussion & conclusion

This work set out to experimentally explore the dynamic couplings present in biorobotic implementations. In doing so, our platform provides a common framework for the disparate fields of biology and robotics to understand and evaluate how they can mutually benefit each other. As Webb shows [Webb, 2006], the transfer of a target biological process into a robotic one is a multi-step operation that can be summed up as: *validating a model of a biological system through a robotic implementation.*

In the case where the robotic implementation is used to better understand a biological process, this validation is clearly necessary to verify that the robotic model can really be used as a replacement for the biological one. The robotic platform can then be employed to make new predictions about the target system in a whole new set of experimental conditions. In the case where the robotic implementation is used to perform the same function as the biological one (biomimetics), the validation is necessary to make sure that the target system's behavior is indeed reproduced.

As we have underlined before, this validation is not straightforward and is prone to misconceptions. For instance, if the artificial system's behavior matches the target system's behavior, it does not necessarily mean that the hypothesis is correct. Conversely, if the artificial system's behavior does *not* match the target system's behavior, it does not either mean that the hypothesis is false [Webb, 2006]! Let us illustrate this concept using the Cyborg Fly.

In our first set of experiments, we showed how a naturalistic feedback does not necessarily lead to a useful response. In the experiments, the transfer functions were chosen so that the robot would perform the fly's intended movements and the fly was given a direct visual feedback from the robot. Nonetheless, the robot failed to perform in any useful way.

The cause of this apparent contradiction lies in the way the biological and robotic systems are abstracted: we implicitly hypothesize that the robot reproduces the fly's behavior. For this, we use models of insect flight control (Equ.(C.3) & (C.4)) and transcribe them into commands that generate similar types of responses in the robot. However, the fly and the robot systems are very different and their responses cannot be so easily compared. The fly turns 90° within 50 ms, the robot takes at least one hundred times longer. The fly uses its wings to advance through a fluid volume while the robot rolls on a flat two-dimensional surface. The claim that "the robot reproduces the fly's behavior" has not taken into account neither the vast spatiotemporal differences nor the differences in implementation that was necessary to compare these disparate systems. In these first experiments, the fly probably saw a

## C.3. DISCUSSION & CONCLUSION

quasi-static image that generated very little optical flow. With the lack of meaningful visual feedback, a robust closed-loop behavior can not be expected.

It is interesting to note that we are fast at finding a critical cause to the failure of our system. However, had the system worked, e.g. due to some artifact, we would have very likely not seen or even intentionally ignored the differences cited above, on the basis that the system was functional. This first set of experiments therefore shows how important it is to correctly make abstraction of the spatiotemporal and implementation differences, and to critically analyze the outcome of a biorobotic implementation.

In the second set of experiments, we showed how we can overcome these differences by altering the model: we link the distance of an object to the expansion speed of a stimulus. By doing so, we compensate for the discrepancy in time constants between the fly and the robot by introducing a derivative operator (position is transformed into speed). This results in a functional closed-loop behavior, where the fly-driven robot is able to avoid obstacles. The underlying hypothesis, however, has changed. We cannot claim anymore to have a naturalistic feedback as we have added an amplification term. This change of hypothesis is often neglected in biorobotic approaches, where the model modifications are seen as small adjustments that help the system run smoothly. As these modifications accumulate, their combined effect can become more important than the biological model itself, strongly undermining the "bio-inspired" aspect.

For the biomimetic case, this results in a non-biomimetic engineered device that is simply ill-named. This does not have a very dramatic effect, except if the optimality of the target biological system is (falsely) used as an argument during the design phase. For the biorobotic case, this can lead to false scientific claims and is therefore quite dangerous, especially if the model modifications are not openly declared and are, as a consequence, very hard to find.

Our third and fourth experiments showed that it is not sufficient for a robotic system to perform just like its biological counterpart to claim that the underlying processes are the same. To illustrate this, we used a partially (C.2.3) or completely (C.2.4) decoupled processes to generate stable closed-loop turning behavior (Equ.C.6). In our case, we have done it intentionally. Unfortunately, such effects are often unforeseen artifacts of the experimental paradigm, and can lead to false claims. For instance, the turning behavior of fruit flies in tethered setups is strongly affected by the lack of gyroscopic feedback. It is therefore dangerous to make conclusions on free-flight turning based on tethered experiments.

The simple "success/failure" classification done here has the advantage of providing clear, unambiguous, results for our different case studies. Our future work will nonetheless complement this qualitative assessment, and characterize quantitatively the performance of the Cyborg system.

In summary, the biorobotic platform presented here offers a unique way to analyze the couplings between biology and robotics. Our work has demonstrated how the choice of coupling between the biology and robotic processes often lead to unexpected properties. From these observations, the guidelines to a meaningful pursuit of biorobotic approaches can be drawn. Such understanding is crucial to biomimetic implementations and robotic-augmented biology. It may also potentially benefit the medical field of brain-robot interfaces.

# References

[Abbott, 2006] Abbott, A. (2006). Neuroprosthetics: In search of the sixth sense. *Nature*, 442(7099):125–127.

[Abbott, 2007] Abbott, A. (2007). Biological robotics: Working out the bugs. *Nature*, 445(7125):250–253.

[Abbott et al., 2007] Abbott, J. J., Nagy, Z., Beyeler, F., and Nelson, B. J. (2007). Robotics in the small, part I: Microbotics. *IEEE Robotics and Automation Magazine*, 14(2):92–103.

[Allen et al., 1993] Allen, P. K., Timcenko, A., Yoshimi, B., and Michelman, P. (1993). Automated tracking and grasping of a moving object with a robotic hand eye system. *IEEE Transactions on Robotics and Automation*, 9(2):152–165.

[Balint and Dickinson, 2001] Balint, C. N. and Dickinson, M. H. (2001). The correlation between wing kinematics and steering muscle activity in the blowfly Calliphora vicina. *Journal of Experimental Biology*, 204(24):4213–26.

[Balint and Dickinson, 2004] Balint, C. N. and Dickinson, M. H. (2004). Neuromuscular control of aerodynamic forces and moments in the blowfly, Calliphora vicina. *Journal of Experimental Biology*, 207(22):3813–3838.

[Bartussek et al., prep] Bartussek, J., Shchekinova, E., Graetzel, C. F., Saleh, H., Howard, J., Zapotocky, M., and Fry, S. ((in prep.)).

[Bender and Dickinson, 2006a] Bender, J. A. and Dickinson, M. H. (2006a). A comparison of visual and haltere-mediated feedback in the control of body saccades in Drosophila melanogaster. *Journal of Experimental Biology*, 209:4597–4606.

[Bender and Dickinson, 2006b] Bender, J. A. and Dickinson, M. H. (2006b). Visual stimulation of saccades in magnetically tethered Drosophila. *Journal of Experimental Biology*, 209(16):3170–3182.

[Beyeler et al., 2008] Beyeler, F., Muntwyler, S., Nagy, Z., Graetzel, C. F., Moser, M., and Nelson, B. J. (2008). Design and calibration of a mems sensor for measuring force and torque acting on a magnetic microrobot. *Journal of Micromechanics and Microengineering*, 18:7.

[Birch and Dickinson, 2001] Birch, J. M. and Dickinson, M. H. (2001). Spanwise flow and the attachment of the leading-edge vortex on insect wings. *Nature*, 412(6848):729–733.

[Borst, 1986] Borst, A. (1986). Time course of the houseflies' landing response. *Biological Cybernetics*, 54(6):379–383.

[Borst and Egelhaaf, 1989] Borst, A. and Egelhaaf, M. (1989). Principles of visual motion detection. *Trends in Neurosciences*, 12(8):297–306.

[Buchner, 1984] Buchner, E. (1984). Behavioral analysis of spatial vision in insects. In Ali, M., editor, *Photoreception and vision in invertebrates*, pages 561–621. Plenum Press, New York.

[Buckholz, 1981] Buckholz, R. H. (1981). Measurements of unsteady periodic forces generated by the blowfly flying in a wind-tunnel. *Journal of Experimental Biology*, 90(Feb):163–173.

[Cembrano et al., 2004] Cembrano, G. L., Rodriguez-Vazquez, A., Galan, R. C., Jimenez-Garrido, F., Espejo, S., and Dominguez-Castro, R. (2004). A 1000 FPS at 128 x 128 vision processor with 8-Bit digitized I/O. *IEEE Journal of Solid-State Circuits*, 39(7):1044–1055.

[Cloupeau et al., 1979] Cloupeau, M., Devillers, J. F., and Devezeaux, D. (1979). Direct measurements of instantaneous lift in desert locust - comparison with jensen experiments on detached wings. *Journal of Experimental Biology*, 80(Jun):1–15.

[Collett et al., 1993] Collett, T., Nalbach, H. O., and Wagner, H. (1993). Visual stabilization in arthropods. *Reviews of Oculomotor Research*, 5:239–63.

[Datteri and Tamburrini, 2007] Datteri, E. and Tamburrini, G. (2007). Biorobotic experiments for the discovery of biological mechanisms. *Philosophy of Science*, 74(3):409–430.

[Dickinson, 2006] Dickinson, M. (2006). Insect flight. *Current Biology*, 16(9):309–314.

# REFERENCES

[Dickinson, 1999] Dickinson, M. H. (1999). Haltere-mediated equilibrium reflexes of the fruit fly, Drosophila melanogaster. *Philosophical Transactions of the Royal Society of London. Series B: Biological Sciences*, 354(1385):903–916.

[Dickinson et al., 2000] Dickinson, M. H., Farley, C. T., Full, R. J., Koehl, M. A. R., Kram, R., and Lehman, S. (2000). How animals move: An integrative view. *Science*, 288(5463):100–106.

[Dickinson and Gotz, 1996] Dickinson, M. H. and Gotz, K. G. (1996). The wake dynamics and flight forces of the fruit fly Drosophila melanogaster. *Journal of Experimental Biology*, 199(9):2085–2104.

[Dickinson et al., 1993] Dickinson, M. H., Lehmann, F. O., and Götz, K. G. (1993). The active control of wing rotation by Drosophila. *Journal of Experimental Biology*, 182:173–189.

[Dickinson et al., 1999] Dickinson, M. H., Lehmann, F. O., and Sane, S. P. (1999). Wing rotation and the aerodynamic basis of insect flight. *Science*, 284(5422):1954–1960.

[Eckert, 1973] Eckert, H. (1973). Optomotorische Untersuchungen am visuellen System der Stubenfliege Musca domestica L. *Biological Cybernetics*, 14(1):23.

[Ellington, 1984a] Ellington, C. (1984a). The aerodynamics of hovering insect flight: I. the quasi-steady analysis. *Philosophical Transactions of the Royal Society of London. Series B: Biological Sciences*, 305:1–16.

[Ellington, 1984b] Ellington, C. P. (1984b). The aerodynamics of hovering insect flight .2. morphological parameters. *Philosophical Transactions of the Royal Society of London Series B-Biological Sciences*, 305(1122):17–40.

[Ellington, 1984c] Ellington, C. P. (1984c). The aerodynamics of hovering insect flight .6. lift and power requirements. *Philosophical Transactions of the Royal Society of London Series B-Biological Sciences*, 305(1122):145–181.

[Fahlbusch et al., 2002] Fahlbusch, S., Shirinov, A., and Fatikow, S. (2002). Afm-based micro force sensor and haptic interface for a nanohandling robot. In *IEEE International Conference on Intelligent Robots and System*, volume 2, pages 1772–1777.

[Fayyazuddin and Dickinson, 1996] Fayyazuddin, A. and Dickinson, M. H. (1996). Haltere afferents provide direct, electrotonic input to a steering motor neuron in the blowfly, calliphora. *Journal of Neuroscience*, 16(16):5225–5232. Article. 0270-6474 English.

[Franceschini et al., 2007] Franceschini, N., Ruffier, F., and Serres, J. (2007). A bio-inspired flying robot sheds light on insect piloting abilities. *Current Biology*, 17(4):329–335.

[Fry et al., 2004] Fry, S. N., Muller, P., Baumann, H. J., Straw, A. D., Bichsel, M., and Robert, D. (2004). Context-dependent stimulus presentation to freely moving animals in 3D. *Journal of Neuroscience Methods*, 135(1-2):149–157.

[Fry et al., 2008] Fry, S. N., Rohrseitz, N., Straw, A. D., and Dickinson, M. H. (2008). Trackfly: Virtual reality for a behavioral system analysis in free-flying fruit flies. *Journal of Neuroscience Methods*, 171(1):110–117.

[Fry et al., 2003] Fry, S. N., Sayaman, R., and Dickinson, M. H. (2003). The aerodynamics of free-flight maneuvers in Drosophila. *Science*, 300(5618):495–498.

[Fry et al., 2005] Fry, S. N., Sayaman, R., and Dickinson, M. H. (2005). The aerodynamics of hovering flight in Drosophila. *Journal of Experimental Biology*, 208(12):2303–2318.

[Frye and Dickinson, 2004] Frye, M. A. and Dickinson, M. H. (2004). Closing the loop between neurobiology and flight behavior in Drosophila. *Current Opinion in Neurobiology*, 14(6):729–736.

[Gates, 2007] Gates, B. (2007). A robot in every home. *Scientific American*, 296(1):58–65.

[Gibbs et al., 2004] Gibbs, M. R. J., Hill, E. W., and Wright, P. J. (2004). Magnetic materials for MEMS applications. *Journal of Physics D-Applied Physics*, 37(22):237–244.

[Gotz, 1968] Gotz, K. G. (1968). Flight control in drosophila by visual perception of motion. *Kybernetik*, 4(6):199.

[Götz, 1964] Götz, K. (1964). Optomotorische Untersuchung des visuellen Systems einiger Augenmutanten der Fruchtfliege Drosophila. *Kybernetik*, 2:77–92.

[Gunter et al., 2007] Gunter, C., Cesari, F., Nath, D., Chou, I. H., Eccleston, A., and Dhand, R. (2007). Genome labours bear fruit. *Nature*, 450(7167):183–183.

[Guzella, 2007] Guzella, L. (2007). *Analysis and Synthesis of Single-Input Single-Output Control Systems*. VDF Hochschulverlag AG.

[Halloy et al., 2007] Halloy, J., Sempo, G., Caprari, G., Rivault, C., Asadpour, M., Tache, F., Said, I., Durier, V., Canonge, S., Ame, J. M., Detrain, C., Correll, N., Martinoli, A., Mondada, F., Siegwart, R., and Deneubourg, J. L. (2007). Social integration of robots into groups of cockroaches to control self-organized choices. *Science*, 318(5853):1155–1158.

[Hassenstein and Reichardt, 1956] Hassenstein, B. and Reichardt, W. (1956). Systemtheoretische Analyse der Zeit-, Reihenfolgen- und Vorzeichenauswertung bei der Bewegungsperzeption des Rüsselkafers Chlorophanus. *Zeitschrift für Naturforschung*, 11b:513–524.

[Hausen, 1982] Hausen, K. (1982). Motion sensitive interneurons in the optomotor system of the fly. 1. the horizontal cells: structure and signals. *Biological Cybernetics*, 45(2):143–156.

[Heide and Gotz, 1996] Heide, G. and Gotz, K. G. (1996). Optomotor control of course and altitude in drosophila melanogaster is correlated with distinct activities of at least three pairs of flight steering muscles. *Journal of Experimental Biology*, 199(8):1711–1726.

[Heisenberg, 1977] Heisenberg, B. (1977). The role of retinula cell types in visual behavior of Drosophila melanogaster. *Journal of Comparative Physiology A*, pages 127–162.

[Heisenberg and Wolf, 1979] Heisenberg, M. and Wolf, R. (1979). On the fine-structure of yaw torque in visual flight orientation of Drosophila-melanogaster. *Journal of Comparative Physiology*, 130(2):113–130.

[Heisenberg and Wolf, 1988] Heisenberg, M. and Wolf, R. (1988). Reafferent control of optomotor yaw torque in drosophila-melanogaster. *Journal of Comparative Physiology a-Sensory Neural and Behavioral Physiology*, 163(3):373–388.

[Hertz, 2008] Hertz, G. (2008). Cockroach controlled mobile robot.

[Hesselberg and Lehmann, 2007] Hesselberg, T. and Lehmann, F.-O. (2007). Turning behaviour depends on frictional damping in the fruit fly Drosophila. *Journal of Experimental Biology*, 210(24):4319–4334.

[Imai et al., 2004] Imai, Y., Namiki, A., Hashimoto, K., and Ishikawa, M. (2004). Dynamic active catching using a high-speed multifingered hand and a high-speed vision system. In *IEEE International Conference on Robotics and Automation*, volume 2, pages 1849–1854.

[Jensen, 1956] Jensen, M. (1956). Biology and physics of locust flight. III. the aerodynamics of locust flight. *Philosophical Transactions of the Royal Society of London Series B-Biological Sciences*, 239(667):511–552.

[Joesch et al., 2008] Joesch, M., Plett, J., Borst, A., and Reiff, D. F. (2008). Response properties of motion-sensitive visual interneurons in the lobula plate of Drosophila melanogaster. *Current Biology*, 18(5):368–374.

[Juusola and Hardie, 2001] Juusola, M. and Hardie, R. C. (2001). Light adaptation in Drosophila photoreceptors: I. Response dynamics and signaling efficiency at 25 degrees C. *Journal of General Physiology*, 117(1):3–25.

[Kagami et al., 2006] Kagami, S., Komuro, T., Watanabe, Y., and Ishikawa, M. (2006). A real-time vision system using a digital vision chip. *Electronics and Communications in Japan Part II-Electronics*, 89(6):34–43.

[Kennedy, 1940] Kennedy, J. (1940). The visual responses of flying mosquitoes. *Proceedings of the Zoological Society of London (A)*, 109(221):42.

[Kositsky et al., 2003] Kositsky, M., Karniel, A., Alford, S., M., F. K., and A., M.-I. F. (2003). Dynamical dimension of a hybrid neurorobotic system. *IEEE Transactions on Neural Systems and Rehabilitation Engineering*, 11(2):155–159.

[Krapp and Hengstenberg, 1997] Krapp, H. G. and Hengstenberg, R. (1997). A fast stimulus procedure to determine local receptive field properties of motion-sensitive visual interneurons. *Vision Research*, 37(2):225–234.

[Kuhn, 1962] Kuhn, T. (1962). *The Structure of Scientific Revolutions*.

[Kunze, 1961] Kunze, P. (1961). Untersuchung des bewegungssehens fixiert fliegender bienen. *Journal of Comparative Physiology*, 44(6):656.

# REFERENCES

[Lehmann, 2001] Lehmann, F. O. (2001). The efficiency of aerodynamic force production in Drosophila. *Comparative Biochemistry and Physiology A*, 131(1):77–88.

[Lehmann and Dickinson, 1997] Lehmann, F. O. and Dickinson, M. H. (1997). The changes in power requirements and muscle efficiency during elevated force production in the fruit fly Drosophila melanogaster. *Journal of Experimental Biology*, 200(7):1133–1143.

[Lichtsteiner et al., 2006] Lichtsteiner, P., Posch, C., and Delbruck, T. (2006). A 128 x 128 120dB 30mW asynchronous vision sensor that responds to relative intensity change. In *IEEE International Conference on Solid-State Circuits*, pages 2060–2069.

[Lindemann et al., 2003] Lindemann, J. P., Kern, R., Michaelis, C., Meyer, P., van Hateren, J. H., and Egelhaaf, M. (2003). Flimax, a novel stimulus device for panoramic and highspeed presentation of behaviourally generated optic flow. *Vision Research*, 43(7):779–791.

[Maybeck, 1979] Maybeck, P. (1979). *Stochastic models, estimation and control*, volume 1. Academic Press.

[McRuer and Jex, 1967] McRuer, D. T. and Jex, H. R. (1967). A review of quasilinear pilot models. *IEEE Transactions on Human Factors in Electronics*, 8(3):231–250.

[Miall, 1978] Miall, R. C. (1978). Flicker fusion frequencies of 6 laboratory insects, and response of compound eye to mains fluorescent ripple. *Physiological Entomology*, 3(2):99–106.

[Miao et al., 2007] Miao, W., Lin, Q. Y., and Wu, N. J. (2007). A novel vision chip for high-speed target tracking. *Japanese Journal of Applied Physics Part 1-Regular Papers Brief Communications and Review Papers*, 46(4B):2220–2225.

[Morgan, 1915] Morgan, T. H. (1915). *The mechanism of Mendelian heredity*.

[Muntwyler, 2006] Muntwyler, S. (2006). Design and fabrication of a capacitive 3dof mems force sensor. Technical report, ETH Zürich.

[Nakayama, 1985] Nakayama, K. (1985). Biological image motion processing - a review. *Vision Research*, 25(5):625–660.

# REFERENCES

[Nalbach, 1991] Nalbach, G. (1991). Body and gaze stabilization via sense organs for rotational velocity: analysis of the haltere function with vibrational stimuli. *Verh Deutsch Zool Ges*, 84:355.

[Nalbach G, 1986] Nalbach G, H. R. (1986). Die halteren von Calliphora als drehsinnesorgan. *Verh Dtsch Zool Ges*, 79.

[Nasir et al., 2005] Nasir, M., Dickinson, M., and Liepmann, D. (2005). Multidirectional force and torque sensor for insect flight research. volume 1, pages 555–558 Vol. 1.

[Ogawa et al., 2005] Ogawa, N., Oku, H., Hashimoto, K., and Ishikawa, M. (2005). Microrobotic visual control of motile cells using high-speed tracking system. *IEEE Transactions on Robotics*, 21(4):21.

[Oku et al., 2005] Oku, H., Ogawa, N., Ishikawa, M., and Hashimoto, K. (2005). Two-dimensional tracking of a motile micro-organism allowing high-resolution observation with various imaging techniques. *Review of Scientific Instruments*, 76(3):34301.

[Purcell, 1977] Purcell, E. M. (1977). Life at low reynolds-number. *American Journal of Physics*, 45(1):3–11.

[Rafal Zbikowski and Knowles, 2006] Rafal Zbikowski, S. A. A. and Knowles, K. (2006). Topical review: On mathematical modelling of insect flight dynamics in the context of micro air vehicles. *Bioinspiration & Biomimetics*, 1(2):26–37.

[Ramamurti and Sandberg, 2007] Ramamurti, R. and Sandberg, W. C. (2007). A computational investigation of the three-dimensional unsteady aerodynamics of Drosophila hovering and maneuvering. *Journal of Experimental Biology*, 210(5):881–896.

[Rangelow et al., 2002] Rangelow, I. W., Grabiec, P., Gotszalk, T., and Edinger, K. (2002). Piezoresistive SXM sensors. *Surface and Interface Analysis*, 33(2):59–64.

[Rao et al., 1995] Rao, B., Gao, R., and Friedrich, C. (1995). Integrated force measurement for on-line cutting geometry inspection. *Instrumentation and Measurement, IEEE Transactions on*, 44(5):977–980.

[Reichardt, 1961] Reichardt, W. (1961). Autocorrelation, a principle for relative movement discrimination by the central nervous system. In Rosenblith, W., editor, *Sensory communication*, pages 303–317. MIT Press, New York.

# REFERENCES

[Reiser and Dickinson, 2008] Reiser, M. B. and Dickinson, M. H. (2008). A modular display system for insect behavioral neuroscience. *Journal of Neuroscience Methods*, 167(2):127–139.

[Reynolds and Riley, 2002] Reynolds, D. R. and Riley, J. R. (2002). Remote-sensing, telemetric and computer-based technologies for investigating insect movement: a survey of existing and potential techniques. *Computers and Electronics in Agriculture*, 35(2-3):271–307.

[Rohrseitz and Fry, ] Rohrseitz, N. and Fry, S. In prep.

[Sane, 2003] Sane, S. P. (2003). The aerodynamics of insect flight. *Journal of Experimental Biology*, 206(23):4191–4208.

[Sane and Dickinson, 2001] Sane, S. P. and Dickinson, M. H. (2001). The control of flight force by a flapping wing: Lift and drag production. *Journal of Experimental Biology*, 204(15):2607–2626.

[Sane and Dickinson, 2002] Sane, S. P. and Dickinson, M. H. (2002). The aerodynamic effects of wing rotation and a revised quasi-steady model of flapping flight. *Journal of Experimental Biology*, 205(8):1087–1096.

[Sato et al., 2008] Sato, H., Berry, C., Casey, B., Lavella, G., Yao, Y., VandenBrooks, J., and Maharbiz, M. (2008). A cyborg beetle: Insect flight control through an implantable, tetherless microsystem. In *MEMS 2008*, Tucson, AZ, USA.

[Schuurman and Capson, 2004] Schuurman, D. C. and Capson, D. W. (2004). Robust direct visual servo using network-synchronized cameras. *IEEE Transactions on Robotics and Automation*, 20(2):319–334.

[Sedov, 1965] Sedov, L. (1965). *Two-Dimensional Problems in Hydrodynamics and Aerodynamics*.

[Sherman and Dickinson, 2003] Sherman, A. and Dickinson, M. H. (2003). A comparison of visual and haltere-mediated equilibrium reflexes in the fruit fly Drosophila melanogaster. *Journal of Experimental Biology*, 206(2):295–302.

[Strauss et al., 1997] Strauss, R., Schuster, S., and Gotz, K. G. (1997). Processing of artificial visual feedback in the walking fruit fly drosophila melanogaster. *Journal of Experimental Biology*, 200(9):1281–1296.

[Sun and Lan, 2004] Sun, M. and Lan, S. L. (2004). A computational study of the aerodynamic forces and power requirements of dragonfly (aeschna juncea) hovering. *Journal of Experimental Biology*, 207(11):1887–1901.

[Sun et al., 2005] Sun, Y., Fry, S. N., Potasek, D. P., Bell, D. J., and Nelson, B. J. (2005). Characterizing fruit fly flight behavior using a microforce sensor with a new comb-drive configuration. *Journal of Microelectromechanical Systems*, 14(1):4–11.

[Sun and Nelson, 2004] Sun, Y. and Nelson, B. (2004). Mems for cellular force measurements and molecular detection. *International Journal of Information Acquisition (IJIA)*, 1:23–32.

[Sun et al., 2003] Sun, Y., Potasek, D., Piyabongkarn, D., Rajamani, R., and Nelson, B. (2003). Actively servoed multi-axis microforce sensors. In *IEEE International Conference on Robotics and Automation, 2003*, volume 1, pages 294–299.

[Tammero and Dickinson, 2002] Tammero, L. F. and Dickinson, M. H. (2002). Collision-avoidance and landing responses are mediated by separate pathways in the fruit fly, Drosophila melanogaster. *Journal of Experimental Biology*, 205(18):2785–2798.

[Tanaka and Kawachi, 2006] Tanaka, K. and Kawachi, K. (2006). Response characteristics of visual altitude control system in Bombus terrestris. *Journal of Experimental Biology*, 209(22):4533–4545.

[Taylor et al., 2008] Taylor, G. K., Bacic, M., Bomphrey, R. J., Carruthers, A. C., Gillies, J., Walker, S. M., and Thomas, A. L. R. (2008). New experimental approaches to the biology of flight control systems. *Journal of Experimental Biology*, 211(2):258–266.

[Taylor and Thomas, 2003] Taylor, G. K. and Thomas, A. L. R. (2003). Dynamic flight stability in the desert locust Schistocerca gregaria. *Journal of Experimental Biology*, 206(16):2803–2829.

[Taylor and Zbikowski, 2005] Taylor, G. K. and Zbikowski, R. (2005). Nonlinear time-periodic models of the longitudinal flight dynamics of desert locusts Schistocerca gregaria. *Journal of the Royal Society Interface*, 2(3):197–221.

[Von Holst and Mittlestaedt, 1950] Von Holst, E. and Mittlestaedt, H. (1950). Das Reafferenzprinzip. *Naturwissenschaften*, 37(20):464–476.

# REFERENCES

[Wang et al., 2004] Wang, Z. J., Birch, J. M., and Dickinson, M. H. (2004). Unsteady forces and flows in low reynolds number hovering flight: two-dimensional computations vs robotic wing experiments. *Journal of Experimental Biology*, 207(3):449–460.

[Webb, 2006] Webb, B. (2006). Validating biorobotic models. *Journal of Neural Engineering*, 3(3):R25–R35.

[Wehner, 1981] Wehner, R. (1981). Spatial vision in arthropods. In *Handbook of Sensory Physiology*, volume VII/6C, pages 287–616. Springer, Berlin, Heidelberg, New York, Tokyo.

[Weisfogh and Jensen, 1956] Weisfogh, T. and Jensen, M. (1956). Biology and physics of locust flight .1. basic principles in insect flight - a critical review. *Philosophical Transactions of the Royal Society of London Series B-Biological Sciences*, 239(667):415–458.

[Weiss et al., 1987] Weiss, L. E., Sanderson, A. C., and Neuman, C. P. (1987). Dynamic sensor-based control of robots with visual feedback. *IEEE Journal of Robotics and Automation*, 3(5):404–417.

[Wessberg et al., 2000] Wessberg, J., Stambaugh, C. R., Kralik, J. D., Beck, P. D., Laubach, M., Chapin, J. K., Kim, J., Biggs, J., Srinivasan, M. A., and Nicolelis, M. A. L. (2000). Real-time prediction of hand trajectory by ensembles of cortical neurons in primates. *Nature*, 408(6810):361–365.

[Wilburn et al., 2005] Wilburn, B., Joshi, N., Vaish, V., Talvala, E. V., Antunez, E., Barth, A., Adams, A., Horowitz, M., and Levoy, M. (2005). High performance imaging using large camera arrays. *ACM Transactions on Graphics*, 24(3):765–776.

[Wood, 2008] Wood, R. J. (2008). The first takeoff of a biologically inspired at-scale robotic insect. *IEEE Transactions on Robotics*, 24(2):341–347.

[Woods et al., 2001] Woods, M. I., Henderson, J. F., and Lock, G. D. (2001). Energy requirements for the flight of micro air vehicles. *Aeronautical Journal*, 105(1045):135–149.

[Yesin et al., 2006] Yesin, K. B., Vollmers, K., and Nelson, B. J. (2006). Modeling and control of untethered biomicrorobots in a fluidic environment using electromagnetic fields. *International Journal of Robotics Research*, 25(5-6):527–536.

[Zaagman et al., 1977] Zaagman, W., Mastebroek, H., Buyse, T., and Kuiper, J. (1977). Receptive field characteristics of a directionally selective movement detector in the visual system of the blow fly. *Journal of Comparative Physiology A*, 116(1):39.

[Zanker and Gotz, 1990] Zanker, J. M. and Gotz, K. G. (1990). The wing beat of Drosophila-melanogaster .2. dynamics. *Philosophical Transactions of the Royal Society of London Series B-Biological Sciences*, 327(1238):19–44.

[Zhang and Liu, 2006] Zhang, G. B. and Liu, J. (2006). High time-resolution visual motion detection with time stamped pixel design. *Analog Integrated Circuits and Signal Processing*, 46(2):153–158.

Die VDM Verlagsservicegesellschaft sucht für wissenschaftliche Verlage abgeschlossene und herausragende

## Dissertationen, Habilitationen, Diplomarbeiten, Master Theses, Magisterarbeiten usw.

für die kostenlose Publikation als Fachbuch.

Sie verfügen über eine Arbeit, die hohen inhaltlichen und formalen Ansprüchen genügt, und haben Interesse an einer honorarvergüteten Publikation?

Dann senden Sie bitte erste Informationen über sich und Ihre Arbeit per Email an *info@vdm-vsg.de*.

**Sie erhalten kurzfristig unser Feedback!**

VDM Verlagsservicegesellschaft mbH
Dudweiler Landstr. 99
D - 66123 Saarbrücken

Telefon +49 681 3720 174
Fax     +49 681 3720 1749

**www.vdm-vsg.de**

Die VDM Verlagsservicegesellschaft mbH vertritt

Printed by Books on Demand GmbH, Norderstedt / Germany